高等学校计算机教材

Android 实用教程

(基于 Android Studio·含视频分析)

郑阿奇　主编

电子工业出版社
Publishing House of Electronics Industry
北京·BEIJING

内容简介

本书以 Android Studio 2.x 作为平台，系统介绍 Android 平台 APP 开发。以应用开发为主线分类介绍 Android Studio 的功能、控件、属性、事件和方法，不过多地说明系统生成的代码细节，而是在应用中理解主要的和基本的内容及其相互关系。配套的实例分为两种类型，一种是知识性的，一种是应用性的。应用性的实例既是独立的，后面又会配套使用，使后面的内容介绍既有一定的规模和应用感觉，又只需要说明当下内容。

本书包括习题和实验，习题是消化知识，实验是完成应用性的实例。二维码扫码视频主要结合开发环境分析应用实例。本书提供教学课件、实例工程文件、配套安装文件包和视频文件，可从华信教育资源网（www.hxedu.com.cn）免费获取。

本书可作为大学本科和高职高专院校有关专业的教材和教学参考书，也可作为 Android 自学用书和开发参考。

未经许可，不得以任何方式复制或抄袭本书之部分或全部内容。
版权所有，侵权必究。

图书在版编目（CIP）数据

Android 实用教程：基于 Android Studio · 含视频分析/郑阿奇主编. —北京：电子工业出版社，2017.6
ISBN 978-7-121-31883-2

Ⅰ.①A… Ⅱ.①郑… Ⅲ.①移动终端－应用程序－程序设计－高等学校－教材 Ⅳ.①TN929.53

中国版本图书馆 CIP 数据核字（2017）第 130990 号

策划编辑：程超群
责任编辑：裴 杰
印　　刷：三河市鑫金马印装有限公司
装　　订：三河市鑫金马印装有限公司
出版发行：电子工业出版社
　　　　　北京市海淀区万寿路 173 信箱　邮编　100036
开　　本：787×1092　1/16　印张：20　字数：512 千字
版　　次：2017 年 6 月第 1 版
印　　次：2019 年 2 月第 5 次印刷
定　　价：45.00 元

凡所购买电子工业出版社图书有缺损问题，请向购买书店调换。若书店售缺，请与本社发行部联系，联系及邮购电话：（010）88254888，88258888。
质量投诉请发邮件至 zlts@phei.com.cn，盗版侵权举报请发邮件至 dbqq@phei.com.cn。
本书咨询联系方式：（010）88254577，ccq@phei.com.cn。

前　言

如今，Android 已是全世界市场占有率和普及率最高的智能手机操作系统，在中国大陆更是家喻户晓，其中文名称为"安卓系统"。正因为如此，近年来，基于 Android 的 APP 开发是市场的热点之一。

传统的 Android 开发使用 Eclipse，它是一个开放源代码、基于 Java 的可扩展开发平台。Eclipse 只是一个框架和一组服务，用于通过插件组件构建开发环境。要开发 Android 程序，就必须安装专为 Android 平台定制的 ADT 插件。在 Eclipse 上安装 Android 开发环境，首先需要安装支持 Java 应用程序运行的 Java 开发工具包 JDK，然后再按顺序安装 Eclipse、Android SDK 和 ADT。由于 ADT 本身并未集成 Android 的 API，需要用户根据实际情况自己上网选择更新所需 API 的 SDK 版本，常常会出现安装的开发环境中各组件之间的兼容性问题，搭建这样的环境对开发者的要求较高，初学者往往不易上手。

2013 年 5 月 16 日，Google 推出新的 Android 开发环境——Android Studio，它基于 IntelliJ IDEA，类似 Eclipse ADT，但它的内部集成了 Android SDK，并提供了集成的 Android 开发工具用于开发和调试，无论安装还是使用都比 Eclipse + ADT 方式要方便得多。目前最新版本为 Android Studio 2.2。

本书以 Android Studio 2.x 作为平台，系统介绍 Android 平台 APP 开发，包括 Android 概述、Android 开发入门、Android 用户界面、用户界面布局、Android 多页面与版块、Android 用户界面进阶、Android 服务与广播程序设计、Android 数据存储与共享以及 Android 地图应用开发。介绍 Android 以应用开发为主线，分类介绍 Android Studio 的功能、控件、属性、事件和方法，不过多地说明系统生成的代码细节，而是在应用中理解主要的和基本的内容及其相互关系。配套的实例分为 2 种类型，一种是知识性的，一种是应用性的。应用性的实例既是独立的，后面又会配套使用，使后面的内容介绍既有一定的规模和应用感觉，又只需要说明当下内容。

本书每章包含二维码扫码视频，通过实例把主要内容联系起来讲解，分析文件关系、代码之间的相互联系，介绍解决问题过程和要点，回答读者关心的问题。

本书提供配套的教学课件、全部应用实例代码（工程文件）、构建与教材相同开发和实验的平台需要的安装文件包以及本书视频文件，需要者可从华信教育资源网（www.hxedu.com.cn）免费下载。

本书包括习题和实验，习题是消化知识，实验是完成应用性的实例。

本书由郑阿奇（南京师范大学）主编，参加本书编写的还有曹弋、徐文胜、丁有和、殷红先、陈瀚、陈冬霞、邓拼搏、高茜、刘博宇、彭作民、钱晓军、孙德荣、陶卫冬、吴明祥、王志瑞、徐斌、俞琰、严大牛、郑进、周何骏、于金彬、马骏、周怡明、姜乃松、梁敬东、陆文周等。

由于我们的水平有限，错误在所难免，敬请广大师生、读者批评指正。

意见建议邮箱：easybooks@163.com

编　者

本书配套工程和视频目录

配套工程和视频	页码	配套工程和视频	页码
2.1.1HelloWorld	13	5.3.3 例五 Fragment（分类预览图书-生命周期）	156
2.2.1HelloWorld1（增加图片和命令按钮）	32	6.1.1 例一 OptionMenu（选项菜单）	159
2.2.3HelloWorld2（增加命令按钮事件处理）	40	6.1.1 例一 OptionMenubyCode（代码定义选项菜单）	163
3.2.1 例 3.1TextTest（文本框使用）	51	6.1.2 例二 SubMenu（子菜单）	164
3.2.2 例 3.2ButtomTest（按钮使用）	53	6.1.2 例二 SubMenubyCode（代码定义子菜单）	168
3.2.3 例 3.3CheckBoxTest（复选框使用）	56	6.1.3 例三 BookPage（快捷菜单）	170
3.2.4 例 3.4RadioTest（单选按钮使用）	58	6.1.4 例四 ActionBar（操作栏）	178
3.2.5 例 3.5SpinnerTest（下拉列表使用）	60	6.2 例五 DetailFragment（Fragment 页面）	180
3.2.6 例 3.6ListViewTest（列表框使用）	62	6.3.2 例六 TabHost（Tab 导航栏应用）	190
3.4.1 例 3.7WebViewTest（网页浏览控件使用）	72	7.2.2 例 7.1CircleStartService（启动方式使用 Service）	199
3.4.2 例 3.8HorizontalScrollViewTest（滚动预览控件）	77	7.2.3 例 7.2CircleBindService（绑定方式使用 Service）	204
3.4.3 例 3.9ImageSwitcherTest（照片查看器）	80	7.2.4 例 7.3PowerMonitorNormal（多 Service 交互）	210
3.4.4 例 3.10BarTest（拖动与等级条）	83	7.3.2 例 7.4PowerMonitorNormalBroadcast（普通广播）	221
4.1.1 例 4.1LinearLayoutTest（线性布局）	90	7.3.3 例 7.5SmsFilterOrderedBroadcast（有序广播）	223
4.1.2 例 4.2RelativeLayoutTest（相对布局）	94	8.1.2 例一 SharedReg/SharedLog（共享优先存储）	229
4.1.3 例 4.3TableLayoutTest（表格布局）	97	8.2.2 例二 InterReg（文件存储）	239
4.1.4 例 4.4GridLayoutTest（网格布局）	102	8.3.2 例三 SqliteReg（SQLite 应用）	243
4.2.1 例一 LoginPage（登录界面）	107	8.4.3 例四 SqliteReg/SqliteLog（ContentProvider 共享）	251
4.2.2 例二 RegisterPage（注册界面）	111	9.2 例 9.1MyBd（设置地图类型及区域检索）	267
4.2.3 例三 BookPage（图书展示）	116	9.3 例 9.2MyBd（地理经纬度检索）	274
5.1.2 例一 LoginPage（登录响应）	126	9.4 例 9.3MyBd（Poi 检索）	279
5.1.3 例二 RegisterPage（注册成功直接登录）	132	9.5 例 9.4MyBd（驾驶路径规划）	286
5.2.3 例三 LoginPage（登录响应-生命周期）	142	9.6 例 9.5MyBd（公交线路查询）	295
5.3.2 例四 Fragment（分类预览图书）	147		

目 录

第1章 Android 概述 ... 1
1.1 Android 简介 ... 1
1.2 Android 开发平台 ... 3
1.3 Android Studio 2.x 的安装 ... 4
1.3.1 在 Windows 上安装 Java 开发工具包 ... 4
1.3.2 安装 Android Studio ... 8

第2章 Android 开发入门 ... 13
2.1 创建 Android 工程 ... 13
2.1.1 第一个 Android 工程：HelloWorld ... 13
2.1.2 Android Studio 工程开发环境 ... 16
2.1.3 Android Studio 工程结构 ... 22
2.1.4 模拟运行 ... 26
2.1.5 真机运行 ... 30
2.2 修改 HelloWorld 程序 ... 32
2.2.1 可视化修改界面 ... 32
2.2.2 配置界面文本 ... 37
2.2.3 代码编写与事件处理 ... 40
2.3 升级 Android Studio 工程 ... 46

第3章 Android 用户界面 ... 48
3.1 用户界面基础 ... 48
3.1.1 用户界面基本要求 ... 48
3.1.2 控件概述 ... 49
3.2 基本的界面控件 ... 51
3.2.1 字符显示和编辑控件：TextView/EditText ... 51
3.2.2 按钮和图像按钮控件：Button/ImageButton ... 53
3.2.3 复选框：CheckBox ... 56
3.2.4 单选按钮及其容器：RadioButton 和 RadioGroup ... 58
3.2.5 下拉列表：Spinner ... 60
3.2.6 列表框：ListView ... 62
3.3 界面事件 ... 63
3.3.1 按键事件 ... 63
3.3.2 触摸事件 ... 67
3.4 高级控件应用 ... 72
3.4.1 网页浏览控件：WebView ... 72
3.4.2 滚动预览控件：HorizontalScrollView ... 77
3.4.3 照片查看器：ImageSwitcher ... 80
3.4.4 条类控制器：SeekBar/RatingBar ... 83

第4章 用户界面布局 ... 88
4.1 界面布局 ... 88
4.1.1 线性布局：LinearLayout ... 90
4.1.2 相对布局：RelativeLayout ... 94
4.1.3 表格布局：TableLayout ... 97
4.1.4 网格布局：GridLayout ... 102
4.1.5 绝对布局：AbsoluteLayout ... 106
4.1.6 版块布局：FrameLayout ... 107
4.2 用户界面综合实例 ... 107
4.2.1 【例一】：登录界面 ... 107
4.2.2 【例二】：注册界面 ... 111
4.2.3 【例三】：图书展示 ... 116

第5章 Android 多页面与版块 ... 123
5.1 Intent 页面间数据传递 ... 123
5.1.1 Intent 原理 ... 123
5.1.2 基本数据类型传递方式（【例一】：登录响应） ... 126
5.1.3 对象数据类型传递方式（【例二】：注册成功直接登录） ... 132
5.2 Activity 生命周期 ... 140
5.2.1 Activity 概述 ... 140
5.2.2 生命周期的基本概念 ... 140
5.2.3 Activity 的生命周期（【例三】：登录响应-生命周期） ... 142
5.3 Fragment（页面版块） ... 145
5.3.1 Fragment 的生命周期 ... 145
5.3.2 Fragment 应用（【例四】：分类预览图书） ... 147
5.3.3 Fragment 生命周期（【例五】：分类预览图书-生命周期） ... 156

第6章 Android 用户界面进阶 ... 159
6.1 菜单 ... 159
6.1.1 选项菜单（【例一】：调用第4章例二、例三和第5章例一） ... 159
6.1.2 子菜单（【例二】：第4章例二、例三组和第5章例一分类组） ... 164
6.1.3 快捷菜单（【例三】：根据第4章例三选择图书显示详细信息） ... 170
6.1.4 操作栏（【例四】：实现例二分组菜单） ... 178
6.2 Fragment 页面（【例五】：图书列表和详细信息不同页和同页显示） ... 180
6.3 Tab 导航栏 ... 189
6.3.1 Tab 导航栏介绍 ... 189
6.3.2 Tab 导航栏应用（【例六】：实现例二分组菜单） ... 190

第7章 Android 服务与广播程序设计 ... 195
7.1 Java 线程编程基础 ... 195
7.2 Service（服务）程序设计 ... 196
7.2.1 Service 概述 ... 196
7.2.2 启动方式使用 Service ... 199
7.2.3 绑定方式使用 Service ... 204

		7.2.4 多 Service 交互及生命周期 ……………………………… 210

- 7.3 广播（BroadcastReceiver） ……………………………… 218
  - 7.3.1 BroadcastReceiver 概述 ……………………………… 218
  - 7.3.2 普通广播应用 ……………………………… 221
  - 7.3.3 有序广播应用 ……………………………… 223

第 8 章 Android 数据存储与共享 ……………………………… 228

- 8.1 SharedPreferences（共享优先）存储 ……………………………… 228
  - 8.1.1 SharedPreferences 概述 ……………………………… 228
  - 8.1.2 SharedPreferences 应用（【例一】：存取注册信息） ……………………………… 229
- 8.2 内部文件存储 ……………………………… 237
  - 8.2.1 Android 系统文件访问 ……………………………… 238
  - 8.2.2 文件存储应用（【例二】：存取注册信息） ……………………………… 239
- 8.3 SQLite 数据库存储与共享 ……………………………… 242
  - 8.3.1 SQLite 概述 ……………………………… 242
  - 8.3.2 SQLite 应用（【例三】：存取注册信息） ……………………………… 243
- 8.4 ContentProvider 数据共享组件 ……………………………… 247
  - 8.4.1 ContentProvider 组件 ……………………………… 247
  - 8.4.2 ContentProvider 创建 ……………………………… 249
  - 8.4.3 ContentProvider 应用（【例四】：获取注册信息） ……………………………… 251

第 9 章 Android 地图应用开发 ……………………………… 256

- 9.1 创建地图开发环境 ……………………………… 256
  - 9.1.1 百度地图环境 ……………………………… 256
  - 9.1.2 高德地图环境 ……………………………… 262
- 9.2 设置地图类型及区域检索 ……………………………… 267
  - 9.2.1 设计界面 ……………………………… 267
  - 9.2.2 功能实现 ……………………………… 270
  - 9.2.3 运行效果 ……………………………… 273
- 9.3 地理经纬度检索 ……………………………… 274
  - 9.3.1 设计界面 ……………………………… 274
  - 9.3.2 功能实现 ……………………………… 276
  - 9.3.3 运行效果 ……………………………… 278
- 9.4 Poi 检索 ……………………………… 279
  - 9.4.1 添加类库 ……………………………… 279
  - 9.4.2 设计界面 ……………………………… 279
  - 9.4.3 功能实现 ……………………………… 282
  - 9.4.4 运行效果 ……………………………… 285
- 9.5 驾驶路径规划 ……………………………… 286
  - 9.5.1 添加类库 ……………………………… 286
  - 9.5.2 设计界面 ……………………………… 286
  - 9.5.3 功能实现 ……………………………… 289
  - 9.5.4 运行效果 ……………………………… 295

9.6 公交线路查询 295
    9.6.1 添加类库 295
    9.6.2 设计界面 296
    9.6.3 功能实现 298
    9.6.4 运行效果 301
9.7 高德地图开发 301

习题和实验 304

第 1 章 Android 概述 304
第 2 章 Android 开发入门 305
第 3 章 Android 用户界面 306
第 4 章 用户界面布局 307
第 5 章 Android 多页面与版块 308
第 6 章 Android 用户界面进阶 309
第 7 章 Android 服务与广播程序设计 310
第 8 章 Android 数据存储与共享 311
第 9 章 Android 地图应用开发 311

第 1 章 Android 概述

1.1 Android 简介

1. Android 的起源、发展与现状

Android 是当今最为流行的三大（另外两个为苹果 iOS、微软 Windows Phone）智能手机操作系统之一。Android 一词的本义指"机器人"，最初是由 General Magic 公司软件工程师 Andy Rubin（安迪·鲁宾）开发的一款专用于移动终端设备的 OS（操作系统）软件。2003 年 10 月，Andy Rubin 等人创建 Android 公司，并组建 Android 团队。2005 年 8 月 17 日，Google（谷歌）低调收购了成立仅 22 个月的 Android 公司及其团队。现在被称为 Android 之父的 Andy Rubin 在公司被收购之后成为 Google 公司工程部副总裁，继续负责 Android 项目。而原 Android 这个公司名也就作为 OS 产品的名称被保留了下来。

Google 在收购 Android 公司后，继续对 Android 系统进行开发运营。
- 2007 年 11 月 5 日，Google 正式向外界展示了 Android 手机操作系统，并且在这天宣布建立一个全球性的联盟组织——开放手持设备联盟（Open Handset Alliance）来共同研发和改良 Android 系统。该组织由 84 家手机制造商、软件开发商、电信运营商以及芯片制造商共同组成，以支持 Google 发布的手机操作系统及其上的无数应用软件，Google 则以 Apache 免费开源许可证的授权方式，公开了 Android 的源代码。
- 2008 年 10 月，第一部 Android 智能手机发布。随后，Android 逐渐扩展到平板电脑及其他领域，如电视、数码相机、游戏机等。
- 2011 年第一季度，Android 在全球的市场份额首次超过塞班系统，跃居全球第一。
- 2013 年的第四季度，Android 平台手机的全球市场份额已经达到 78.1%。同年 9 月 24 日，Android 迎来其 5 岁生日，全世界采用这款系统的设备数量已经达到 10 亿台！
- 2014 年第一季度，Android 平台已占所有移动广告流量来源的 42.8%，首度超越了 iOS。

如今 Android 已是全世界市场占有率和普及率最高的智能手机操作系统，最新版本为 Android 7.0。在中国大陆更是家喻户晓，其中文名称为"安卓系统"。

2. Android 的系统架构

Android 和其他操作系统一样，采用了分层的架构，如图 1.1 所示。

从架构图可以看出，Android 系统分为四个层次，从高到低分别是应用程序层（Applications）、应用框架层（Application Framework）、系统运行时库（Libraries/Android Runtime）和 Linux 内核层（Linux Kernel）。

（1）应用程序层

Android 系统一般会与一系列核心应用程序包一起发布，该应用程序包包括客户端、SMS 短消息程序、日历、计算器、地图、浏览器、联系人管理簿等智能手机最常用的应用（App）。所有的应用都是使用 Java 语言编写的。

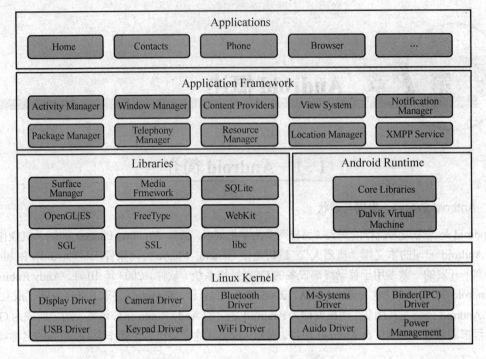

图1.1 Android系统的分层架构

（2）应用框架层

应用框架层是核心应用程序包所使用的API（应用编程接口）框架，应用的开发人员可以完全访问。该API框架的架构设计简化了组件的重用，任何一个应用都可以发布它的功能块并且任何其他的应用也都可以使用其所发布的功能块（不过得遵循框架的安全性）。同样，这种重用机制也使用户能够非常方便地替换程序组件。

隐藏在每个应用后面的是一系列的服务和系统，其中包括以下几方面。

- 活动管理器（Activity Manager）：用来管理应用生命周期并提供常用的导航、回退等功能。
- 内容提供器（Content Providers）：使一个应用可以访问另一个应用的数据（如联系人数据库），或者共享它们自己的数据。
- 视图系统（View System）：可以用来构建应用，它包括列表（Lists）、网格（Grids）、文本框（Text Boxes）、按钮（Buttons），甚至可嵌入Web浏览器。
- 通知管理器（Notification Manager）：使应用程序可以在状态栏中显示自定义的提示信息。
- 资源管理器（Resource Manager）：提供非代码资源的访问，如本地字符串、图形和布局文件（Layout Files）。

（3）系统运行时库

Android包含一些C/C++库，这些库能被Android系统中不同的组件使用。它们通过上层的应用框架为开发者提供服务。以下是一些核心库。

- Surface Manager（门面管理器）：对显示子系统的管理，并且为多个应用提供了2D/3D图层的无缝整合。
- Media Framework（媒体框架）：支持多种常用的音频、视频格式回放和录制，同时支持静态图像文件。编码格式包括MPEG4、H.264、MP3、AAC、AMR、JPG和PNG等。
- SQLite：一款非常流行的嵌入式数据库，支持SQL语言，并且只占很少的内存，尤其适用于手机这类内存受限的设备。
- WebKit：一个最新的 Web 浏览器引擎，支持 Android 浏览器和一个可嵌入的 Web 视图。
- libc（系统 C 库）：一个从 BSD 继承来的标准 C 语言函数库，它是专门为基于 Embedded Linux 的设备定制的。

（4）Linux 内核层

Android 是基于 Linux 的操作系统，其架构的最底层是 Linux 内核，但并不是通常的 GNU/Linux。因为在一般 GNU/Linux 里支持的功能，Android 大都不支持。另外，Google 为了达到商业目的，必须移除被 GNU/GPL 授权证所约束的部分，例如 Android 将驱动程序移到用户空间，使 Linux 的驱动与内核彻底地分开了。

Android 所依赖的 Linux 内核功能包括安全（Security）、存储器管理（Memory Management）、进程管理（Process Management）、网络堆栈（Network Stack）和驱动程序模型（Driver Model）等。

3. Android 平台的优势

（1）开放性

Android 平台的最大优点首先就是其开放性，开放的平台允许任何移动终端厂商加入到 Android 联盟中来。显著的开放性可以使其拥有更多的开发者，随着智能手机的普及和手机应用的日益丰富，Android 平台也很快走向成熟。

开放性对于 Android 的发展而言，有利于积累人气，这里的人气包括消费者和厂商，而对于消费者来讲，最大的受益正是其丰富的软件资源。开放的平台也会带来更激烈的竞争，如此一来，消费者就可以用更低的价位购得心仪的手机。

（2）多硬件兼容

这一点还是与 Android 平台的开放性相关，由于 Android 的开放性，众多的厂商会推出千奇百怪、功能各具特色的多种硬件产品。但硬件设备功能上的差异和特色，却不会影响到数据同步和软件的使用，如从小米手机改用三星，同时还可将小米中优秀的软件也带到三星上使用，联系人等资料更是可以方便地转移。

（3）方便开发

Android 平台提供给第三方开发者一个十分宽泛、自由的环境，不会受到任何条条框框的限制。而且 Android 平台支持二次开发，开发者可以将现成的互联网应用，如百度地图、新浪微博、微信等功能集成进自己的应用，定制出更为强大的功能，可想而知，会有多少新颖别致的软件源源不断地诞生。

1.2 Android 开发平台

1. 主流开发平台

（1）Eclipse + ADT 方式

传统的 Android 开发使用 Eclipse，它是一个开放源代码、基于 Java 的可扩展开发平台。就其本身而言，Eclipse 只是一个框架和一组服务，用于通过插件组件构建开发环境。要开发 Android 程序，就必须安装专为 Android 平台定制的 ADT（Android Development Tools）插件。在 Eclipse 上安装 Android 开发环境，首先需要安装支持 Java 应用程序运行的 Java 开发工具包 JDK（Java Development Kit），然后再按顺序安装 Eclipse、Android SDK 和 ADT。由于 ADT 本身并未集成 Android 的 API，需要用户根据实际情况自己上网选择更新所需 API 的 SDK 版本，常常会出现安装的开发环境中各组件之间的兼容性问题，搭建这样的环境对开发者的要求较高，初学者往往不易上手。

近年来，Eclipse 官方也推出了专门针对 Android 开发的 Eclipse for Android Developers 版本，较之分立的 Eclipse + ADT 环境使用起来要方便一些，有兴趣的读者可以去 Eclipse 官网下载来试用。

（2）Android Studio

2013 年 5 月 16 日，Google 推出新的 Android 开发环境——Android Studio，它基于 IntelliJ IDEA，类似 Eclipse ADT，但它的内部集成了 Android SDK，并提供了集成的 Android 开发工具用于开发和调试，无论是安装还是使用都比 Eclipse + ADT 方式要方便得多。目前最新版本为 Android Studio 2.2.3。

Android Studio 在 2014 年 12 月 8 日正式成为开发 Android 应用程序的官方集成开发环境（Integrated Development Environment，IDE）工具，Android 开发官网也很明确地指出，原先 Eclipse + ADT 的开发模式已经被 Android Studio 取代，建议开发者尽早改用 Android Studio 以获得最好的支持。基于此，本书采用 Android Studio 作为书中 Android 实例程序的开发环境。

2. 软硬件要求

开发 Android 程序对计算机系统的要求较高，Android 各个版本的 SDK 包普遍较大，会占用一定的磁盘空间，且运行 Android 模拟器要耗费大量内存。鉴于此，强烈建议读者学习本书前先升级自己的计算机至较高配置。笔者所用计算机的配置如下。

处理器（CPU）：Intel(R) Core(TM) i5-4200U CPU @ 1.60GHz 2.30GHz，建议至少在 i5 或以上。

内存：8.00GB（7.89GB 可用）。

硬盘：700GB，其中 C 盘 180GB，D 盘 250GB，且 C 盘、D 盘均为高速固态硬盘。

操作系统：64 位 Windows 7 及以上版本。

1.3 Android Studio 2.x 的安装

1.3.1 在 Windows 上安装 Java 开发工具包

Android Studio 使用 Java 工具构建，因此在开始使用 Android Studio 之前，需要确保已经在自己的电脑上安装了 Java 开发工具包（JDK）。

1. 下载 JDK

可以从 Oracle 官网下载到最新版本的 JDK，网址为 http://www.oracle.com/technetwork/java/javase/downloads/index.html，单击页面上的"Java DOWNLOAD"按钮，如图 1.2 所示。

图 1.2 JDK 下载页面

第 1 章　Android 概述

下一步要求选中"Accept License Agreement"单选按钮以接受许可协议。接着，需要选择适合自己操作系统的 JDK。笔者使用的是 64 位 Windows 7，故单击 Windows x64 标签右侧的文件链接"jdk-8u121-windows-x64.exe"，如图 1.3 所示（Oracle 经常会发布 JDK 的更新版本，到本书出版的时候，JDK 应该已经有了更新的版本，因此务必下载最新版），等待安装文件下载完成。下载得到的文件名为"jdk-8u121-windows-x64.exe"，这个文件的大小通常为 200MB 左右，所以下载过程不会花费太长时间。

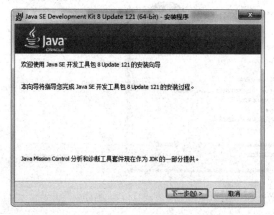

图 1.3　接受许可协议并单击适用于 Windows 的链接

2. 执行 JDK 向导

导航到浏览器下载安装文件的位置，并双击执行该文件。一旦安装开始，将会看到安装向导，如图 1.4 所示。单击"下一步"按钮，系统进入指定安装目录对话框。在 Windows 中，JDK 安装程序的默认路径为 C:\Program Files\Java\。要更改安装目录的位置，可单击"更改"按钮。本书安装到默认路径，参见图 1.5。

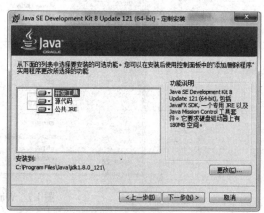

图 1.4　Windows 中的 JDK 安装向导　　　　　图 1.5　选择 JDK 的安装目录

记住安装 JDK 的位置。单击"下一步"按钮，向导提示安装 Java 运行时环境（Java Runtime Environment，JRE）的目标文件夹（C:\Program Files\Java\jre1.8.0_121），默认是与 JDK 相同的目录，同样可单击"更改"按钮选择其他位置。本书保持默认，单击"下一步"按钮，如图 1.6 所示。

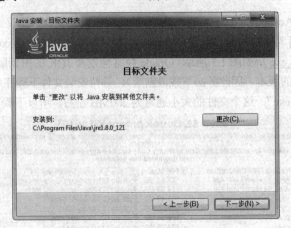

图 1.6 选择 JRE 的目标文件夹

开始安装进程，直到显示安装完成对话框，单击"关闭"按钮，完成安装。

3. 配置 Windows 环境变量

配置 Windows 环境变量的目的是为了让后面安装的 Android Studio 开发平台能够找到 JDK。

在桌面上右击"计算机"图标，从弹出的菜单中选择"属性"，打开"系统"窗口，如图 1.7 所示。

图 1.7 Windows 系统窗口

单击"高级系统设置"选项，系统显示"系统属性"对话框，如图 1.8 所示。单击"环境变量"按钮，系统显示当前环境变量的情况，如图 1.9 所示。

（1）JAVA_HOME 环境变量新建/编辑

在底部列出的"系统变量"列表中，如果 JAVA_HOME 项不存在，单击"新建"按钮创建它。系统显示新建系统变量对话框，在"变量名"文本框中输入 JAVA_HOME，在"变量值"文本框中，输入之前安装 JDK 的位置 C:\Program Files\Java\jdk1.8.0_121，单击"确定"按钮，如图 1.10 所示。

图 1.8　系统属性　　　　　　　　　图 1.9　环境变量

图 1.10　新建 JAVA_HOME 环境变量

以后可以选择 JAVA_HOME 项，单击"编辑"按钮进行修改。
（2）Path 环境变量编辑

Path 环境变量系统已经存在，需要编辑 Path 环境变量，参见图 1.11。将光标置于"变量值"文本框中内容的末尾并输入以下内容：
;C:\Program Files\Java\jdk1.8.0_121\bin
或者
;%JAVA_HOME%\bin

图 1.11　编辑 Path 环境变量

连续三次单击"确定"按钮，Windows 接受这些修改并返回到最初的"系统"窗口。这样，系统就在原来的 Path 路径上增加了一个查找路径。

👀 注意：

在 Windows 10 系统中，编辑 Path 环境变量的界面如图 1.12 所示。

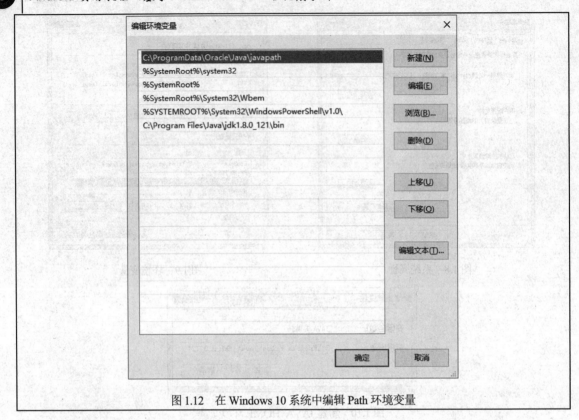

图 1.12　在 Windows 10 系统中编辑 Path 环境变量

（3）测试 JDK 是否正确安装

为了测试新的 JDK 是否已正确安装，单击 Windows 系统的"开始"→"运行"命令，在弹出的对话框中输入 cmd，接着按 Enter 键打开命令行。在命令行窗口中输入以下命令并按 Enter 键：

java -version

如果得到如图 1.13 所示的响应，表示已经正确地安装了 JDK。

图 1.13　确认 JDK 安装正确

1.3.2　安装 Android Studio

对于广大国内用户来说，获取和学习 Android Studio 最便捷的途径是访问 Android 的中文社区（网址：http://www.android-studio.org/），其与 Android 官网（国外）是同步更新的，旨在让 Android 对中文开发者更友好。

1. 下载 Android Studio

在 Android 中文社区首页上，单击"下载 ANDROID STUDIO"按钮开始下载。

> **注意：**
> 下载得到的文件名形如"android-studio-bundle-xxx.xxxxxxx-windows.exe"（其中 x 代表数字），文件大小在 1.6GB 左右（此为集成了 Android SDK 的版本）。我们下载的文件为"android-studio-bundle-145.3537739-windows.exe"。

如果只有几百兆，说明其中不含 Android SDK，用户还要单独去下载 Android SDK 来安装，随之而来的各种兼容性问题将十分棘手！故强烈建议初学者要下载使用集成好 Android SDK 的 Android Studio 版本。

2. 执行安装向导

下载完毕后，双击执行得到的文件，启动安装向导。注意，因为在 Android Studio 的安装过程中需要时刻到网络获得所需的各种文件，为了防止出现麻烦，建议关闭 Windows 防火墙。

在安装向导启动后，单击"Next"按钮向前推进界面，一直到达"Choose Components"界面。在这里，选中所有组件复选框，如图 1.14 所示。

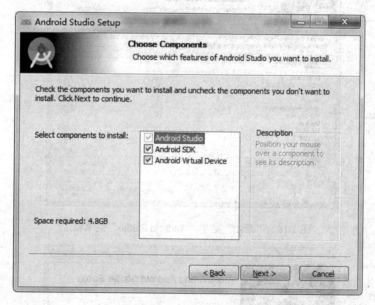

图 1.14 选择组件

单击"Next"按钮，继续向前推进安装界面。在"LicenseAgree"对话框单击"I Agree"按钮接受各种条款。当到达"Configuration Settings Install Locations"界面时，如图 1.15 所示，选择 Android Studio 和 Android SDK 的安装位置，本书采用默认。

单击"Next"按钮，系统显示"Choose Start Menu Folder"对话框，让用户选择是否在系统的开始菜单中创建"Android Studio"快捷菜单项。这里选择默认创建快捷菜单项。

如图 1.16 所示，单击"Install"按钮，系统开始安装，直到安装完成，显示"Installation Complete"。单击"Next"按钮，到达"Completing Android Studio Setup"界面，如图 1.17 所示。"Start Android Studio"复选框能够让 Android Studio 在单击"Finish"之后启动。确保选中了该复选框，接着继续单击"Finish"，Android Studio 将会启动。此后，将需要通过桌面图标或开始菜单来启动 Android Studio。

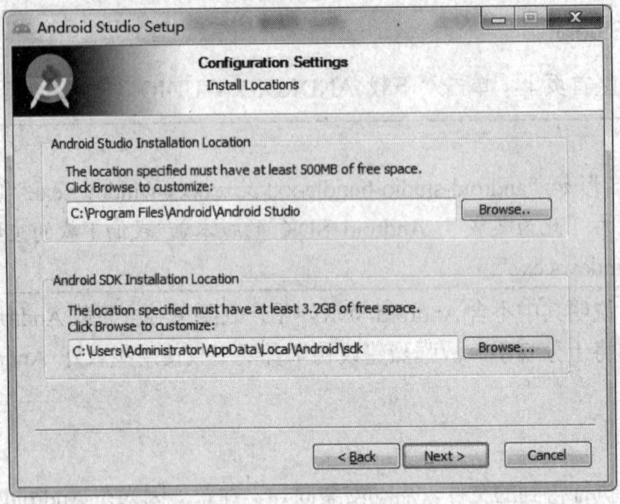

图 1.15 为 Android Studio 和 SDK 选择安装位置

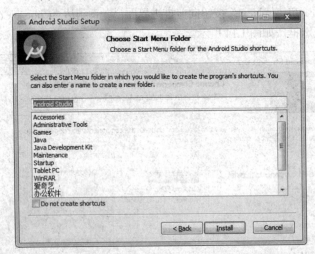

图 1.16 "开始"菜单"Android Studio"菜单项

图 1.17 完成 Android Studio 的安装

3. 第一次启动

当 Android Studio 第一次启动时，它会检查用户的系统之前是否安装过早期版本，并询问用户是否要导入先前版本 Android Studio 的设置，如图 1.18 所示。一般初学者建议使用初始设置，选中下面一个单选按钮后，单击"OK"按钮。

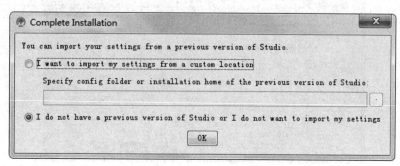

图 1.18　Android Studio 的初始设置

接着出现启动画面，首先弹出的是 Android Studio Setup Wizard 的 Welcome（欢迎）界面，如图 1.19 所示。Setup Wizard 将会分析用户的系统，查找已有 JDK（例如之前安装的那个）以及 Android SDK 的位置。单击"Next"按钮。

图 1.19　Setup Wizard 欢迎界面

在接下来的对话框中，选择安装类型为 Standard（标准），如图 1.20 所示。单击"Next"按钮。

在最终确认界面上，汇总显示了开发环境将要下载安装的全部 Android SDK 组件的详细信息，如图 1.21 所示。单击"Finish"按钮。

Setup Wizard 会下载在 Android Studio 中开发应用需要的所有组件，如图 1.22 所示。稍等一会儿，待完成后单击"Finish"按钮，关闭 Setup Wizard。

至此，Android Studio 开发环境全部安装完成，可以使用了。

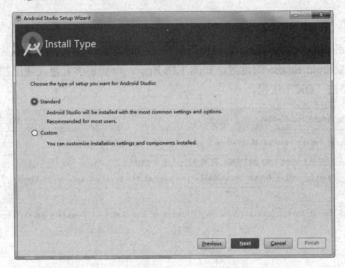

图 1.20 选择安装类型

图 1.21 将要下载安装的 Android SDK 组件

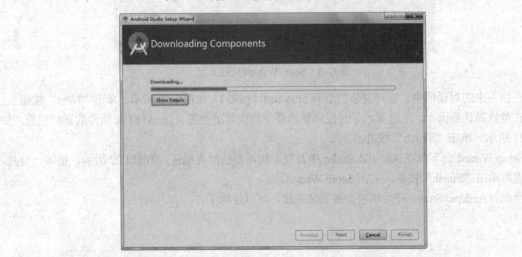

图 1.22 Setup Wizard 下载组件

第 2 章 Android 开发入门

2.1 创建 Android 工程

2.1.1 第一个 Android 工程：HelloWorld

要开发一个 Android 应用（Application，简称 APP）必须先创建 Android 工程（project，又称为项目）。开发一个 APP 需要用到许多资源，这些资源通过一个 project 进行管理。

创建 Android 工程的步骤如下。

（1）启动 Android Studio 后出现如图 2.1 所示窗口，点击"Start a new Android Studio project"来创建新的 Android 工程。

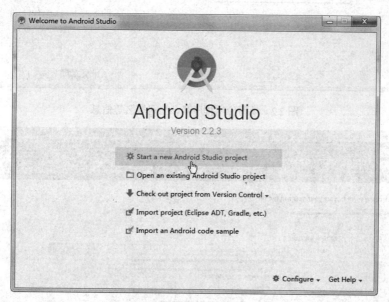

图 2.1 创建一个新的 Android Studio 工程

（2）在"New Project"页填写应用程序名等相关的信息，如图 2.2 所示。填写完单击"Next"按钮进入下一步。

- Application name：应用程序名（也就是工程的名称），默认为"My Application"，这里我们改为"HelloWorld"。
- Company Domain：公司域名，默认为"administrator.example.com"，这里改为"android.easybooks.com"。
- Package name：工程中 Java 程序所在的包名称，由 Android Studio 根据上述 Application name 与 Company Domain 自动生成，也可以通过单击右侧"Edit"进行修改。

● Project location：工程所存放的路径，可以修改。

（3）在"Target Android Devices"页选择要开发的平台与 API 版本，如图 2.3 所示。因为要开发手机或平板电脑的应用程序，所以勾选"Phone and Tablet"；Minimum SDK 设置为"API 15: Android 4.0.3 (IceCreamSandwich)"，代表要安装此应用程序的目标设备，其操作系统最低要求为 Android 4.0.3。单击"Next"按钮进入下一步。

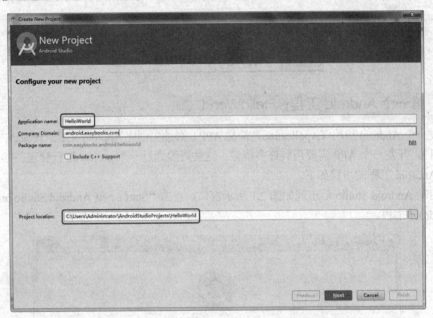

图 2.2 填写应用程序名、公司域名等信息

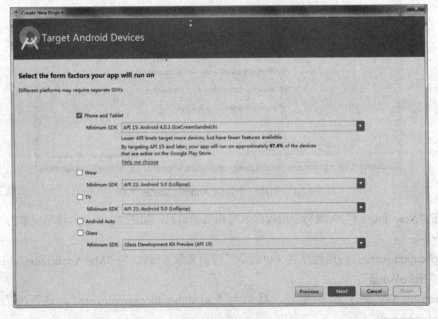

图 2.3 选择应用程序要运行的平台

（4）在"Add an Activity to Mobile"页选择 Activity（画面控制器）类型，如图 2.4 所示。建议选

用"Basic Activity"最基本的类型。单击"Next"按钮进入下一步。

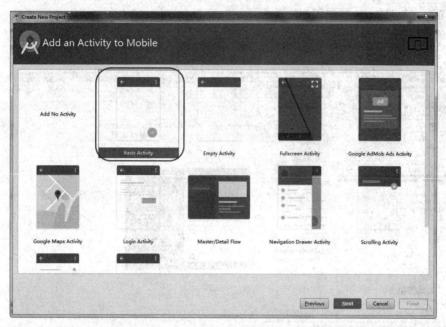

图 2.4　选择 Activity 类型

（5）在"Customize the Activity"页填写定制活动页的相关信息，如图 2.5 所示，填写完毕后单击"Finish"按钮。
- Activity Name：活动页名称。这里为"MainActivity"。
- Layout Name：布局文件名，不可为大写英文字母。这里为"activity_main"。
- Title：标题名称。这里为"MainActivity"。
- Menu Resource Name：菜单资源文件名。这里为"menu_main"。

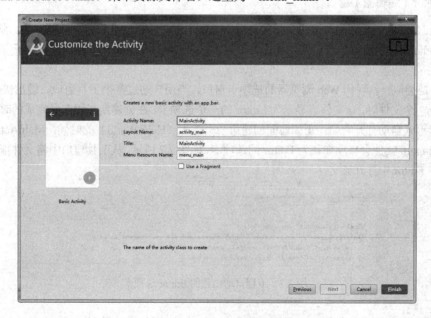

图 2.5　填写 Activity 名称等相关的信息

稍等片刻，在出现"Welcome to Android Studio 2.x"对话框时选择"Close"，系统显示如图2.6所示的开发界面，说明Android工程创建成功。

下面就可以在这个环境中编程开发Android工程了。

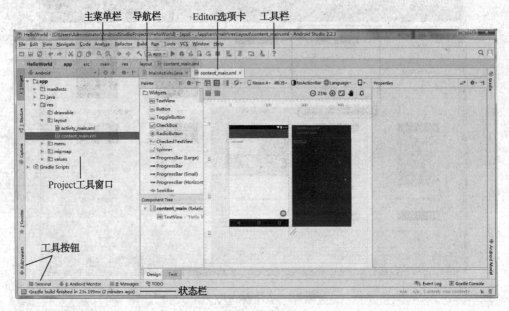

图2.6　Android工程开发界面

2.1.2　Android Studio工程开发环境

Android Studio 窗口环境为 Android 工程开发提供了很方便的手段和方法。注意包括组织工程需要的资源、编辑文件、查找信息、编译运行调试程序。下面继续简要说明。

1. Editor（编辑器）窗口

Android Studio 中允许用户编辑文件的窗口为 Editor 窗口，它位于 IDE 面板的中心。除 Editor 外，Android Studio 中的所有其他窗口则均被称为工具窗口，它们位于 Editor 周围（左侧、下方和右侧）的面板中。

Editor 是类似一款现代 Web 浏览器的选项卡窗口，当用户通过某个工具窗口、键盘快捷键或上下文菜单打开一个文件时，该文件会显示为 Editor 的一个选项卡。在创建新工程向导完成的时候在 Editor 中以选项卡形式自动打开它们。正如前面创建第一个工程（HelloWorld）的时候，MainActivity.java 和 content_main.xml 文件会自动加载为 Editor 的选项卡，如图 2.7 所示。从工具窗口中将文件拖放到 Editor 中，也会在 Editor 中打开该文件。

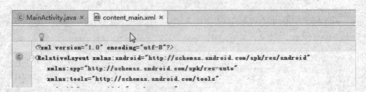

图2.7　工程自动加载的Editor选项卡

其中：

MainActivity.java 文件存放 java 代码，content_main.xml 存放设计界面。

Editor 的上方是 Editor 选项卡，左侧边栏是折叠线，右侧边栏是标记栏。下面逐一介绍各个部分。

（1）Editor 选项卡

要在 Android Studio 的 Editor 选项卡之间进行切换，当然可以用鼠标来选择 Editor 选项卡，还可使用"Alt+向右箭头"或"Alt+向左箭头"。通过主菜单栏的 Window | Editor Tabs 下的菜单项可以进行各种操作。

（2）折叠线

折叠线最显著的特点或许是那些带有颜色的小标签和图标，它们沿着对应的代码行显示，用于指示这些可视化资源，代码展开后两边显示（-），如图 2.8 所示。如果把它们折叠起来，则显示第一行代码，前面加上（+）符号。

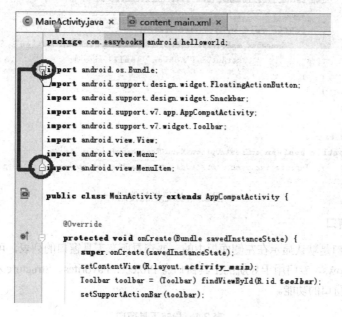

图 2.8　折叠线

折叠线也用于设置断点、完成代码折叠以及显示作用域标识。

（3）标记栏

Editor 的右侧是标记栏，它用于标识源文件中重要行的位置。例如，标记栏会突出显示 Java 或 XML 文件中的警告和编译时的错误。标记栏还会向用户展示未提交的更改、查找结果和书签的位置。

标记栏不会像折叠线那样滚动，相反，标记栏上的颜色标记会出现在文件长度的相对位置。单击标记栏中的颜色标记会立即跳转到文件中的该位置，如图 2.9 所示。

（4）工具按钮

仔细观察 IDE 的左侧、右侧和下方边栏，会发现对应于很多工具窗口的工具按钮，其中一些还标有数字，可以与 Alt 键组合使用来切换该工具按钮对应工具窗口的开启/关闭。

（5）默认布局

Android Studio 默认 IDE 布局是左侧面板中展示 Project 工具窗口，经过一段实际的操作界面感到 IDE 布局有点儿乱时，打开主菜单栏中的 Window|Restore Default Layout，可返回到前面默认 IDE 环境的布局。用户可以定制自己的默认 IDE 布局——打开和关闭有关的工具窗口、设置大小或者摆放它们，然后通过主菜单栏中的 Window|Store Current Layout as Default 将此新布局设置为默认布局。

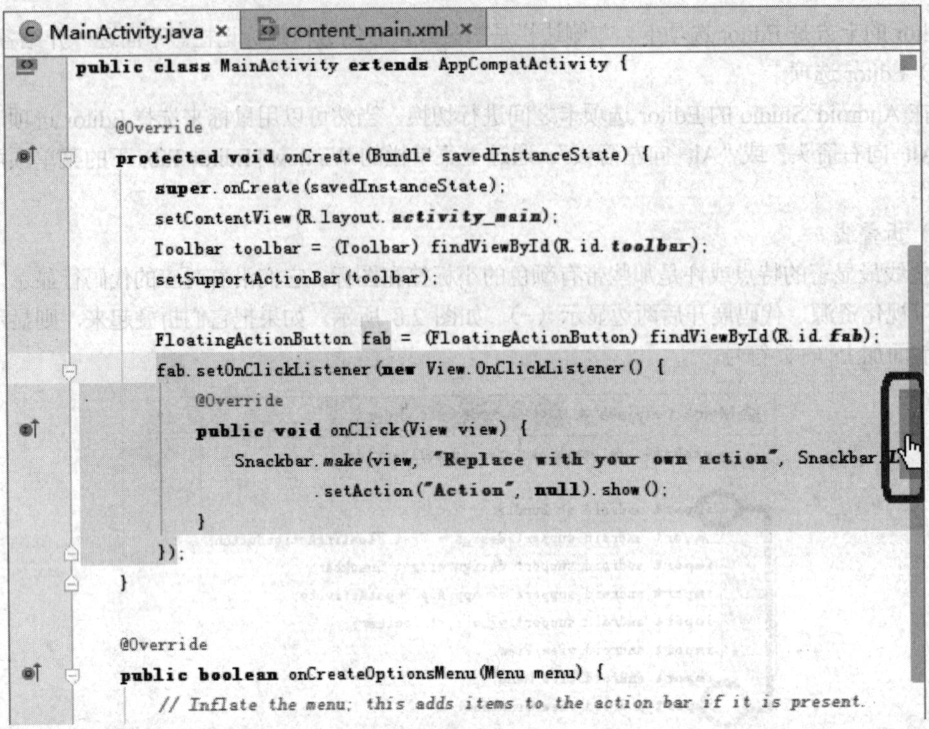

Android 实用教程（基于 Android Studio·含视频分析）

图 2.9 标记栏的使用

2. 导航工具窗口

Project 工具窗口是默认显示在左侧面板中的。要查看所有工具窗口的列表，可以选择主菜单栏中的 View | Tool Windows。专门用于导航的工具窗口有 Project、Favorites、Structure 和 TODO。表 2.1 列出了每个导航工具窗口的功能。

表 2.1 导航工具窗口

工具窗口	快捷键	功能
Project（工程）	Alt+1	允许浏览工程中的文件和资源
Favorites（收藏）	Alt+2	显示收藏、书签和断点
Structure（结构）	Alt+7	显示当前文件中对象或元素的树形结构
TODO		显示工程中所有有效 TODO 的列表

（1）Project 工具窗口

Project 工具窗口是所有导航工具窗口中最有用的一个，该窗口有很多种显示模式（展开窗口左上方的下拉列表可以看到），如图 2.10 所示，其中常用的模式有三种：Project、Packages 和 Android。

Android 和 Project 是最有用的两种模式。在默认情况下，Android Studio 会将模式设置为 Android，但 Android 模式会对用户隐藏某些目录，要想深入领会 Project 工具窗口的功能，需要将窗口的模式设置为 "Project"。Project 工具窗口提供文件和嵌套目录的简单树形界面，可以切换其显示。

Project 工具窗口为用户展示了工程中所有包、目录和文件的概况。如果在 Project 工具窗口中右击某个文件，则会出现一个上下文菜单，其中有三个重要的菜单项："Copy Path"、"File Path" 和 "Show in Explorer"。单击 "Copy Path" 会将此文件在操作系统中的绝对路径复制到剪贴板。单击 "File Path"

会以目录栈的形式显示路径，并以栈顶的文件结束，而单击任意这些目录都会在操作系统中打开它们，如图 2.11 所示。单击 "Show in Explorer" 会在一个新的操作系统窗口中显示该文件。

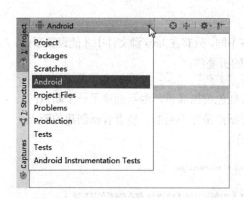

图 2.10　Project 窗口的显示模式

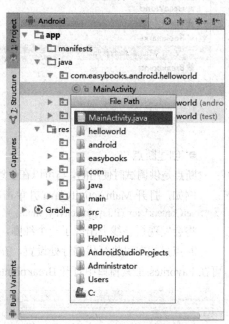

图 2.11　以目录栈形式显示路径

（2）Favorites 工具窗口

当开发 Android 中的功能（或者调试 bug）时，用户可能会创建或修改一些相关的文件。但一个中等复杂度的 Android 工程可能就包含了多达数百个独立文件！因此对相关文件的分组功能就显得很重要。Favorites 工具窗口中包含一些收藏条目，允许用户对相关文件的引用进行逻辑分组，而这些文件可能在物理上位于工程中完全不同的部分。

● 创建收藏条目

收藏条目允许用户立即导航至任意特定的文件或文件组。

例如，在 Editor 选项卡中打开了 MainActivity.java 和 content_main.xml 文件。右击 Editor 中的任意一个选项卡并在上下文菜单中选择 Add All To Favorites | Add All Open Tabs To New Favorites List。在 "Input new favorites list name" 文本框中输入 "main" 并单击 "OK" 按钮，就在 Favorites 工具窗口中创建了一个名为 main 的收藏条目，展开、双击其中列出的某个文件就可以快速地打开/激活它，如图 2.12 所示。

● 创建书签

书签允许用户立即导航至文件中的任意特定行。

例如，将光标置于 MainActivity.java 的任意行中，按 F11 键即可创建或删除任何源文件（包括 XML 文件）中的书签。注意，折叠线中的对号和标记栏中的黑色标记均表示新的书签。可以在 Favorites 工具窗口中查看刚刚创建的书签，也可以随时打开书签，如图 2.13 所示。

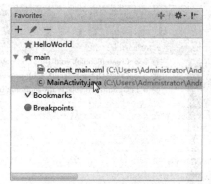

图 2.12　创建的收藏条目

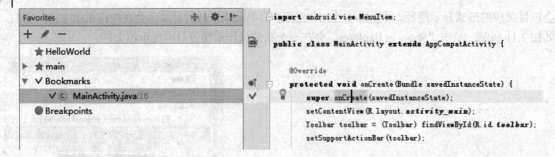

图2.13 创建书签

● 创建断点

断点是供调试时使用的，与可以在任意文件中设置的书签不同，只有在 Java 源文件中才能设置断点。例如，打开 MainActivity.java 并单击靠近以下代码行处的折叠线：

setContentView(R.layout.activity_main);

将会发现折叠线上出现了一个红圈，而且该行也会以淡红色突出显示，表示创建了一个断点。

断点只能在可执行代码行处设置，在注释行上尝试设置断点是行不通的。要查看新创建的断点，可在 Favorites 工具窗口中打开 Breakpoints 树，如图 2.14 所示。

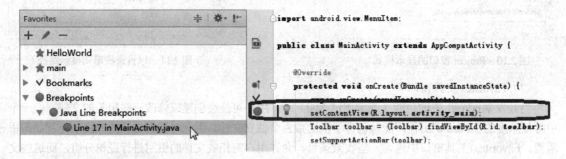

图2.14 创建断点

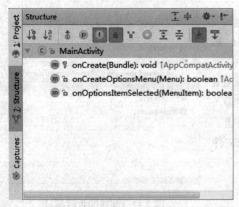

图2.15 Structure 工具窗口

（3）Structure 工具窗口

Structure 工具窗口显示文件中元素的层次结构，单击 Structure 工具窗口中的任意元素会立刻将光标移至该元素在 Editor 中的位置。

例如，当 Editor 显示 Java 源文件（例如 MainActivity.java）时，Structure 工具窗口显示包含字段、方法和内部类等元素的树，如图 2.15 所示。

当 Editor 显示 XML 文件（例如 content_main.xml）时，Structure 工具窗口显示 XML 元素（组件）的树。

（4）TODO 工具窗口

字面意义上"TODO"表示待办事项。TODO 本质上就是注释，用于提醒程序员和他们的合作者还有尚未完成的工作。TODO 的写法类似于注释，以两个向前斜杠开始，单词"TODO"全部大写，然后是一个空格。例如：

//TODO inflate the layout here.

读者可试着在 MainActivity.java 中创建一个 TODO，然后打开 TODO 工具窗口来观察它。在 TODO

工具窗口中单击一个 TODO 会立刻跳转到该 TODO 在源代码中的位置。

3．主菜单栏

主菜单栏位于 Android Studio 的最上方，通过使用它的菜单和子菜单，几乎可以执行任何操作。与 Android Studio 中其他的栏目不同，主菜单栏无法隐藏。主菜单栏及其子菜单项很多，即使最老练的 Android 开发者在日常工作中也仅会用到这些操作中的一小部分，而且这些操作中的大部分均有对应的键盘快捷键和（或）上下文菜单项。

4．工具栏

工具栏中包含频繁使用的文本操作按钮，例如 Cut、Copy、Paste、Undo 和 Redo。工具栏同时包含在 Android Studio 中进行各种管理的按钮，包括 SDK Manager 和 AVD Manager。工具栏还包含用于 Settings 和 Help 的按钮，以及用于 Run 和 Debug 应用的按钮。工具栏中的所有按钮均有对应的菜单项和键盘快捷键。

为了节省屏幕空间，高级用户可以通过取消对 View|Toolbar 菜单项的选中来隐藏工具栏。

5．导航栏

导航栏位于工具栏的下方，外观显示为一个横向的链式箭头框，表示从工程根目录（位于左侧）到 Editor 当前所选选项卡中文件（位于右侧）的路径，如图 2.16 所示。

图 2.16　Android Studio 特有的导航栏

导航栏是 Android Studio 不同于其他 IDE 的一个特色，用于在不借助 Project 工具窗口的情况下，在工程的资源之间进行导航。

6．状态栏

状态栏显示相关且上下文敏感的反馈，例如关于所有正在运行的进程或工程状态的信息。状态栏位于 IDE 界面的底部，其具体构成如图 2.17 所示。

图 2.17　状态栏

（1）切换边栏按钮

状态栏的最左角是切换边栏按钮，单击这个按钮会切换边栏的隐藏和显示。此外，当把鼠标移至此按钮上时，会出现一个上下文菜单，允许用户激活任何工具窗口，如图 2.18 所示。

（2）消息区域

消息区域用于提供反馈并显示关于当前正在运行进程的状态信息。当把鼠标移至菜单项或工具栏中的按钮等 UI（用户界面）元素时，此区域还会显示提示。

（3）编辑器光标位置区域

以"行：列"的格式显示光标在 Editor 中的位置。单击这个区域会激活一个对话框，允许用户直接定位到代码中的某个特定的行。

(4) 行分隔符区域

显示的是文本文件中使用的回车换行格式。在 Windows 上,默认值是 CRLF,表示回车换行符。LF 是 Unix 和 Mac 机器上使用的标准格式,也是 Git 中所采用的。如果正在 Windows 系统上进行开发,Git 通常会在向仓库提交代码时将 CRLF 转换为 LF。

(5) 文本格式区域

描述了用于源文件的文本编码,默认值是 UTF-8,它是 ASCII 的超集且涵盖了大多数西文字符,包括能够在标准 Java 或 XML 文件中找到的所有字符。

(6) 文件访问标识符区域

该区域允许用户在读/写和只读之间进行切换。打开的锁图标表示拥有 Editor 中当前文件的读/写权限。关闭的锁图标则意味着编辑器中的当前文件是只读的。可以通过单击标识符的图标来切换这些设置。

(7) 突出显示级别按钮

该按钮会激活一个带有滑块的对话框,如图 2.19 所示,允许用户设置想要在代码中看到的突出显示级别。

图 2.18 切换边栏按钮

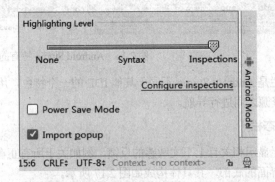

图 2.19 设置突出显示级别

默认设置是"Inspections",它对应一个皱着眉头的检查官大人头像图标。此设置意味着程序员要准备好接受一些严格的检查,因为编译器会严格识别代码中的语法错误和可能的问题(警告)。编程时,用户将会看到在标记栏中以黄色标记的形式生成的一些警告。

滑块的中间设置是"Syntax",在这种设置下,编译器会忽略警告,故 Syntax 模式没有 Inspections 模式严格,但仍然会突出显示那些将会阻止代码编译的语法问题。

滑块最左端的模式是"None",在此模式下,即使是最严重的语法错误也会被忽略,尽管在尝试构建可执行程序的时候,编译器还是会被这些错误卡住。

对于初学者来说,为了避免犯错,推荐将突出显示级别保留为默认的"Inspections",在编写代码阶段就及早地接受编译器的严格检查。

2.1.3 Android Studio 工程结构

一个完整的 Android Studio 工程中包含的文件和资源很多,非常庞杂,但使用 Project 工具窗口在 Android 模式下还是能够比较清楚地看出它的结构。以本章前面创建的 HelloWorld 工程为例,它在

Project 工具窗口中的 Android 模式视图如图 2.20 所示。

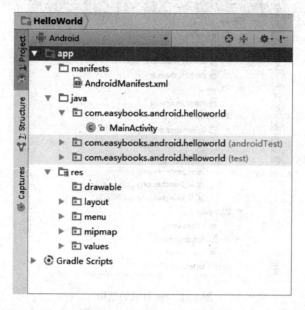

图 2.20 HelloWorld 工程结构

由图可见，一个典型的 Android Studio 工程的目录可分成 3 个部分：manifests 文件、java 目录与 res 目录。

1. manifests 目录

每一个 Android 应用程序都需要 manifests 类型文件，存储着该应用程序的重要信息，系统生成的默认的文件名为 AndroidManifest.xml，初学者一般不需要关注其中的内容。

2. java 目录

java 目录内存放的是工程所有的 Java 源程序文件，Android Studio 工程默认创建的 Java 源文件为 MainActivity.java。文件名是 Android 工程向导根据 ActivityName 自动生成的，初始就由 Editor 窗口打开，初学者一般不需要关注其中的代码。

3. res 目录

工程所需的非程序资源大多放在 res 目录内。其中的文件名只能为小写字母、数字、_（下画线）、.（点）。res 目录的结构如图 2.21 所示。

（1）drawable 目录

drawable 目录存放工程需要的图片文件（png、jpg 等）资源。开发应用时，程序员需将工程要用到的图片资源复制到该目录下，才能在代码中引用，这样设计界面上才能显示出想呈现的图片。

（2）layout 目录

layout 目录专门存放用户界面的布局（layout）文件。选择"Basic Activity"新建 APP 工程时，Android Studio 自动生成了两个文件：activity_main.xml 与 content_main.xml。其中，activity_main.xml 是工程的主显示控制文件，而 content_main.xml 则是主设计文件。初学者一般不用过多关注 activity_main.xml 文件的内容，因为 content_main.xml 才是主设计界面。

Android Studio 支持两种界面设计方式——可视化设计和编写 XML 代码。

Android 实用教程（基于 Android Studio·含视频分析）

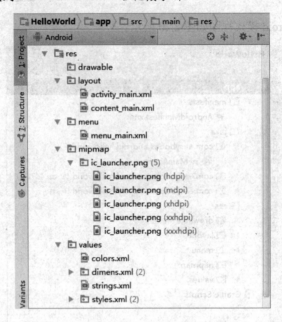

图 2.21 res 目录的结构

第一种方式：可视化设计界面

如图 2.22 所示，当 content_main.xml 文件在 Editor 中处于打开状态时，在 Editor 窗口底部的左下角有两个选项卡："Design"和"Text"，点击"Design"选项卡显示的是可视化设计页，其左侧是一个工具箱面板（Palette），上面有各种带图标的 UI 控件条目，中央区域是一个智能手机屏幕，设计程序界面时用户只需用鼠标选中工具箱内的某个控件，然后将其拖曳到手机屏幕界面上即可。

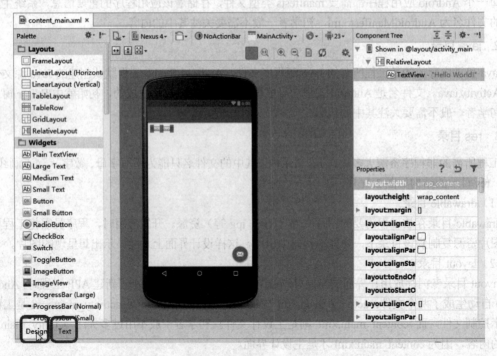

图 2.22 可视化设计方式

页面右上角的"Component Tree"子窗口以树状视图显示了当前设计的程序界面上各控件之间的层次关系。选中手机屏幕上的某个控件，可在页面右下方的"Properties"子窗口中设置其属性。

第二种方式：编写 XML 代码设计界面

也可以直接编写 XML 代码来描述设计用户界面，例如，在 Editor 中编辑"content_main.xml"文件时，点击窗口底部的"Text"选项卡，系统切换到代码视图，如图 2.23 所示。

图 2.23　编写 XML 代码方式

系统初始生成了"<RelativeLayout … >"标签，并在其中包含了一个<TextView … >标签，用于显示"Hello World!"。用户可以在<TextView … >标签处替换成自己需要的标签集，以实现用户需要显示的内容。

（3）menu 目录

menu 只是用于菜单的界面设计，现在手机应用开发并不流行菜单的使用，所以就不做具体介绍了。

（4）mipmap 目录

与 drawable 目录的作用类似，也用于存放工程的图片资源，可以按照屏幕分辨率将图片分成 hdpi（高分辨率）、mdpi（中分辨率）、xhdpi（超高分辨率）、xxhdpi（超超高分辨率）与 xxxhdpi（超超超高分辨率）。在实际开发中，如果希望自己的应用能够同时适应多种不同档次屏幕分辨率的手机，一般建议将相同图片按照分辨率不同而制作成 5 份，分别存放在对应的 5 个目录内。

（5）values 目录

values 目录存放用户界面所需用到的文字（strings.xml）、颜色（colors.xml）和样式（styles.xml）等的定义文件。以文字为例，系统默认 strings.xml 专门存放 Android 应用程序所需用到的文字。虽然文字也可以直接写在 layout 文件内，但不建议这样做。由于手机应用往往要求满足同时支持多国语言，将整个应用的界面文字都集中于 strings.xml 内，翻译人员就只要直接翻译 strings.xml 中的文字即可，而无须到 layout 文件的源码中去寻找要翻译的文字，这样便于应用程序本地化。

4. Android 工程架构信息

单击 Project 工具窗口 Gradle Scripts→build.gradle (Module: app)文件，就会展示如图 2.24 所示的内容，这是本工程的架构信息，可用于查阅，初学者一般不要修改它。

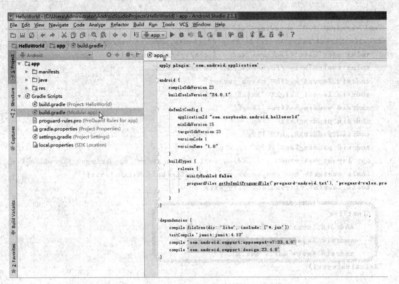

图 2.24　查看工程架构信息

2.1.4　模拟运行

在编写 Android 应用程序过程中需要不断测试程序的结果。为了方便开发人员安装与测试程序，Android Studio 提供了 Android 仿真器（Android Virtual Device，简称 AVD 或称为 emulator）。用仿真器执行 Android 应用程序的方式称为模拟运行，以区别于在真实物理移动设备（如手机、平板电脑）上的真机运行，所以在执行第一个 Android 应用程序前，应该先创建一个 Android 仿真器。

1. 创建 Android 仿真器

创建 Android 仿真器的步骤如下。

（1）在 Android Studio 内，单击主菜单 Tools | Android | AVD Manager，如图 2.25 所示，或在工具栏上单击相同的图标按钮即可开启 AVD Manager（仿真器管理器）。

（2）在"Your Virtual Devices"对话框下单击"+ Create Virtual Device"按钮，会显示可以选择的仿真器硬件类型，如图 2.26 所示，这里我们选一款名为"Nexus 5X"（Google 产品）的手机，单击"Next"按钮继续。

注意，AVD 选择仿真器可以安装的硬件版本，随着 Android SDK 的版本升级，会增加一些可选项。

（3）选择仿真器要安装的操作系统版本。建议选择 ABI 字段值为 x86、Target 带有 Google APIs 的系统映像文件，如图 2.27 所示。x86 代表模拟 Intel x86 Atom CPU，配合 Intel x86 仿真器加速器 1（Android Studio 默认会安装）可以让仿真器执行更顺畅；有 Google APIs 才能执行带有 Google Map 功能的应用程序。单击"Next"按钮继续。

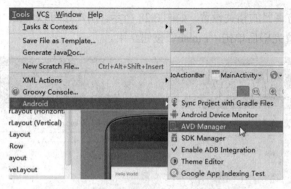

图 2.25 启动 AVD Manager 仿真器管理器

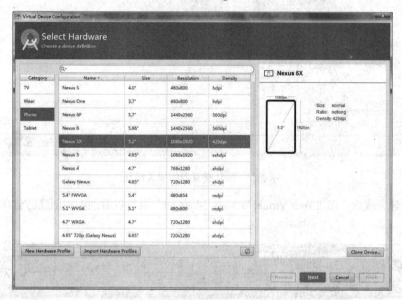

图 2.26 选择可以仿真的硬件类型

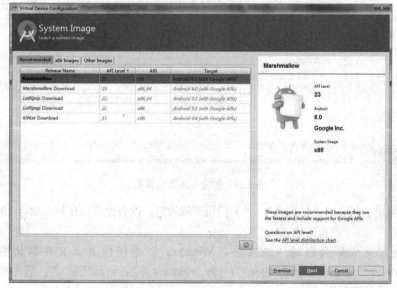

图 2.27 选择好 ABI 字段的值和 Target 系统映像文件

(4) 接下来，确认仿真器的相关设置，如图 2.28 所示。用户还可以单击"Show Advanced Settings"按钮进行高级设置，设置完毕后单击"Finish"按钮结束。

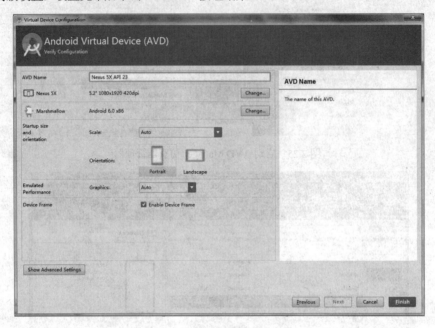

图 2.28　确认仿真器的相关设置

仿真器创建完成后，在"Your Virtual Devices Manager"页可以看到刚刚创建的仿真器，如图 2.29 所示。

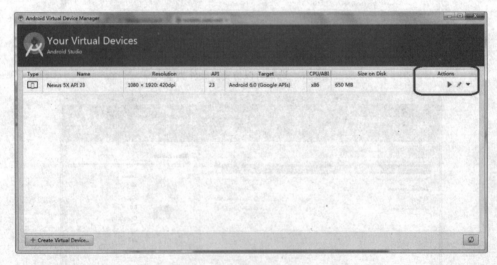

图 2.29　创建完成的仿真器

用户可直接单击其后的"Actions"（▶）按钮来启动它，或者也可以在稍后运行 HelloWorld 程序的时候再启动。

仿真器创建完成后会自动生成文件，Windows 7 系统仿真器文件默认所在的路径为"C:\Users\<user>\.android\avd"，其中<user>为当前 Windows 系统的登录用户名，笔者的计算机是以管理员登录的，故文件路径为"C:\Users\Administrator\.android\avd"。

2. 在仿真器上运行工程

（1）选择仿真器

想在仿真器上运行 Android 工程十分简单，只要单击 Android Studio 主菜单 Run | Run 'app'（或单击工具栏上相同的图标按钮），然后等待片刻就会出现如图 2.30 所示的窗口。直接选中刚刚创建的仿真器 "Nexus 5X API 23"，单击 "OK" 按钮启动它。

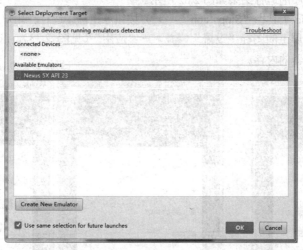

图 2.30　选择要仿真的设备然后在其上运行工程

> **注意：**
> 用户也可以单击 "Create New Emulator" 再创建一个新的仿真器来运行本工程（仿真器的管理和运行都要耗费相当可观的系统资源，除非要用于手机应用的兼容性测试，一般的开发不建议创建多个仿真器），或者勾选 "Use same selection for future launches" 将当前仿真器指定为默认（建议初学者这么做），后面测试运行程序都在它上面进行。

（2）启动仿真器

仿真器启动过程的界面如图 2.31 所示，各类 Android 仿真器运行普遍都要耗费比较多的系统资源，故启动过程也十分漫长！请读者在这一步耐心等待。如果想要获得好一点儿的使用体验，可考虑升级自己的计算机硬件配置以提高性能。

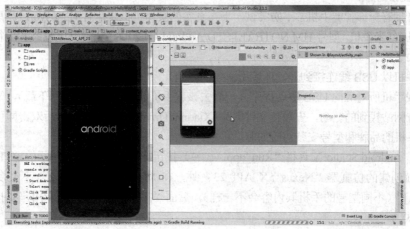

图 2.31　仿真器的启动过程界面

> **注意：**
> 鉴于 Android 仿真器资源耗费量大的现状，为了提高开发效率，建议读者在运行第一个程序启动仿真器后就不要关闭，后面做的程序都直接在这个已启动的仿真器上运行，可节省不少时间。

（3）运行结果

HelloWorld 工程程序的运行结果如图 2.32 所示。

单击 Android Studio 工具栏"Stop 'app'"（■）按钮停止程序运行。用户可通过鼠标点击（模拟手指触击）仿真器手机屏上的图标"设置"（ ⚙ ）→"应用"（见图 2.33），选中"HelloWorld"应用后将其卸载以节省仿真器的内存。

图 2.32 模拟运行 HelloWorld　　　　图 2.33 卸载 HelloWorld 应用

2.1.5 真机运行

如果读者拥有一部智能手机，不妨尝试一下在真实的实体机上来运行 HelloWorld 程序，在此笔者以自己的手机（红米手机，型号 HM NOTE 1LTETD/Android 4.4.2）为例，在其上安装和运行 HelloWorld 工程的步骤如下。

（1）将手机以 USB 线连接到开发环境所使用的电脑。

（2）下载手机所对应的 USB 驱动程序（一般去该款手机品牌生产商的官网下载），笔者就是访问小米手机官方网站找到的驱动，安装包文件名为 Xiaomi_Driver_7.1.exe，解压后双击即可安装。读者请根据自己手机的品牌和型号安装对应的驱动程序。

（3）单击 Android Studio 工具栏"Run 'app'"（▶）按钮，出现如图 2.34 所示的窗口，可以看到，这里除了先前创建的仿真器"Nexus 5X API 23"外，在"Connected Devices"下又多了一个条目"d8585700[...]"（不同型号的手机其内容会不一样），后面"[]"里的英文是提示用户要打开手机上的 USB 调试模式。

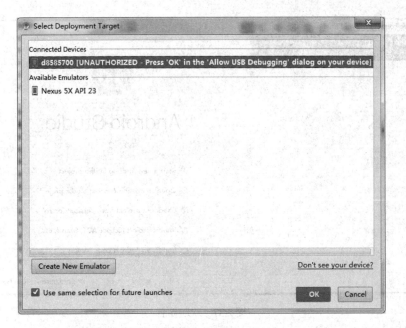

图 2.34 选择 USB 连接的真实设备（手机）运行工程

（4）此时会发现手机屏上弹出一个"允许 USB 调试吗？"通知消息框，手指触击"确定"按钮打开手机的 USB 调试模式，如图 2.35 所示。

（5）回到计算机上如图 2.34 所示的窗口，单击"OK"按钮，同时注意观察手机屏幕，稍等片刻后，HelloWorld 程序就会被编译安装到手机上并自行启动运行，如图 2.36 所示。

图 2.35 打开 USB 调试

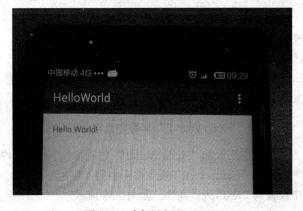

图 2.36 真机运行 HelloWorld

如果初试完成，可以关闭 HelloWorld 工程。选择 File→Close Project。这时，系统显示对话框如图 2.37 所示。

图 2.37 左边显示当前已经操作过的工程，此时用户可以打开已经创建的工程，或者新建工程，或者关闭 Android Studio。

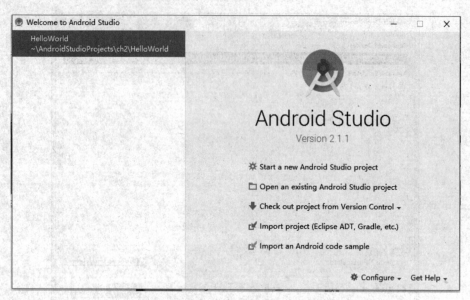

图 2.37 欢迎界面

2.2 修改 HelloWorld 程序

上面运行的 HelloWorld 程序是系统生成的，解决创建一个简单应用程序（工程）的过程。

本节将修改 HelloWorld 程序，使用工具箱面板里的控件来丰富程序的界面元素，并且编写代码实现一些最简单的功能，以进一步让读者熟悉应用程序的过程。

具体要求如下：增加显示一张图片和一个按钮。

2.2.1 可视化修改界面

1. 打开 HelloWorld 工程，进入用户界面设计状态

单击"Project"，选择"content_main.xml"，采用"Design"视图页。

2. 让"Hello World!"字号变大字体变红

（1）原工程的"Hello World!"文字太小，现通过修改 TextView 控件的属性将其放大变醒目。

双击"Component Tree"视图中的"TextView"项（TextView 类型），在下方"Properties"子窗口中设置其"textSize"属性，如图 2.38 所示。

点击"textSize"属性条目后的按钮，弹出如图 2.39 所示的窗口，在其中选择所需的表示字号大小的工程。这里选择 Android Studio 内置的用来表示文字标题行大小的"abc_text_size_headline_material"项，单击"OK"按钮。

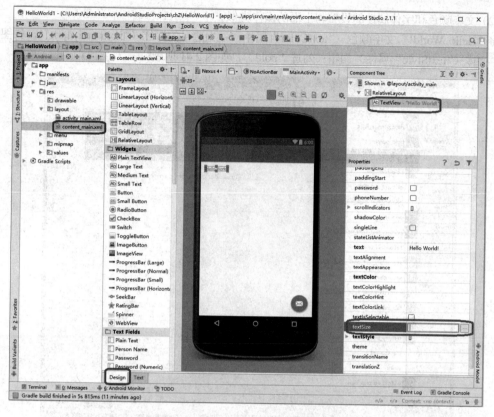

图 2.38　选中 TextView 控件修改其属性

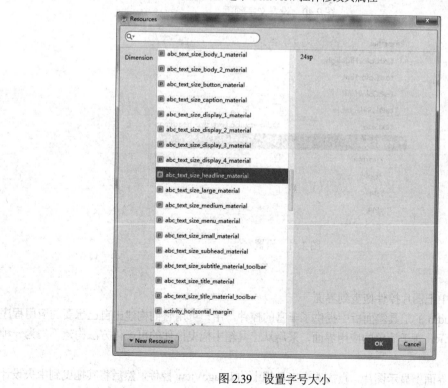

图 2.39　设置字号大小

（2）文字颜色和字体属性的设置与上面的操作类似。

将"Hello World！"文字颜色（"textColor"属性）设为醒目的深红（"holo_red_dark"），如图 2.40 所示。字体（"textStyle"属性）加粗（勾选"bold"），如图 2.41 所示。

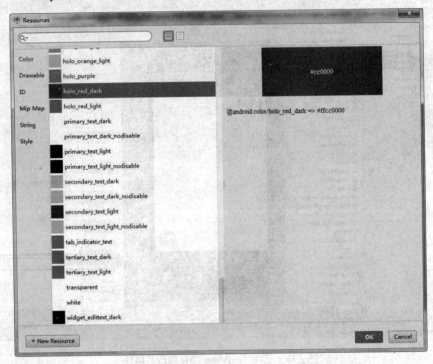

图 2.40　设置文字颜色

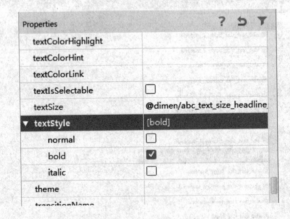

图 2.41　设置字体

3．增加图片

（1）在工具箱中把图片控件拖曳到界面

在 Android Studio 的工具箱面板中提供了丰富的控件，用户可用它们构建出自己想要的应用程序界面，如图 2.42 所示。对于简单的应用界面，采用从工具箱中拖曳控件的可视化方法构建不失为一种有效的方式。

为了能在应用界面上显示图片，点击选中工具箱中的 ImageView 控件，然后将其拖曳到中央设计区的手机屏幕上，如图 2.43 所示。

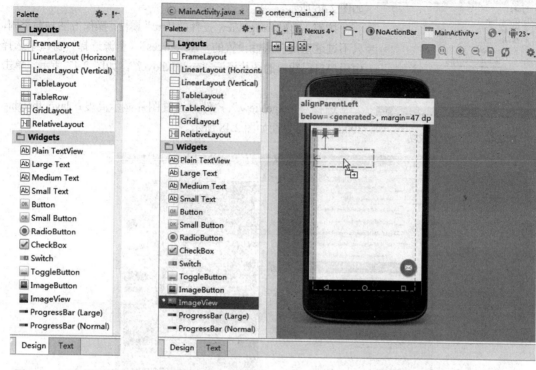

图 2.42　工具箱面板　　　　　　图 2.43　拖曳控件

起初由于尚未设定其显示内容而看不见具体的内容和样子，但 Android Studio 设计模式界面右边 "Component Tree"（组件树）视图可以帮助用户定位、选择已经进入界面的控件，设置任何控件的属性，如图 2.44 所示。其中多了一项 "imageView"——正是刚刚拖曳放置上去的 ImageView 控件。双击该项就会自动将手机屏上对应的控件选中，以便进一步对其进行设计操作。

（2）加载图片资源

要用图像控件 ImageView 显示图片，需要把图片资源加载到工程中。方法是：将需要显示的图片复制到当前工程的 drawable 目录下，如图 2.45 所示。笔者计算机上该目录的路径为 "C:\Users\Administrator\AndroidStudioProjects\HelloWorld\app\src\main\res\drawable"，读者可参考。图片名 "androidlover" 为一个正在读书的可爱小机器人，读者也可使用自己喜欢的其他图片。

图 2.44　"Component Tree"视图　　　　图 2.45　把图片资源放置于特定的目录

图 2.46　图片载入成功

回到开发环境，可在工程 res\drawable 下看到刚刚载入的图片，如图 2.46 所示。

接着，设置 ImageView 控件的"src"属性，操作方法与前类同，只不过这里是在属性设置的"Resources"（资源选择）窗口中选择由用户自己载入的图片资源"androidlover"，如图 2.47 所示，单击"OK"按钮。

选择"ImageView"控件，设置图片显示高度：layoutheight=300dp，屏幕界面如图 2.48 所示。

图 2.47　选择要显示的图片资源

图 2.48　载入图片后的显示效果

4. 加入命令按钮

命令按钮对应的是工具箱面板里的 Button 控件,同样也是采用拖曳的方式拉到手机屏上,如图 2.49 所示。

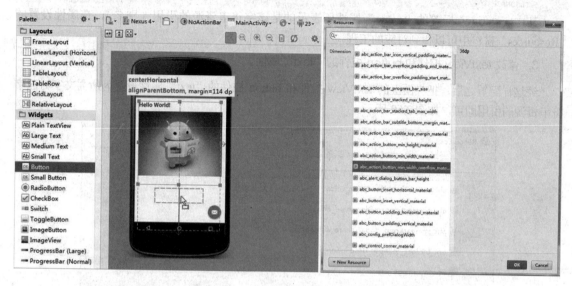

图 2.49 加入按钮控件并调整界面布局

考虑到与"Hello World!"字体大小相配且美观,特将按钮的"textSize"属性设为"abc_action_button_min_width_overflow_material"(36dp),并调整界面布局。

2.2.2 配置界面文本

1. 初始"HelloWorld"工程 content_main.xml 代码

初始生成的 HelloWorld 工程显示的"Hello World!"直接写在程序中作为 TextView 控件的"text"属性存在,代码如下:

```
<?xml version="1.0" encoding="utf-8"?>
<RelativeLayout xmlns:android="http://schemas.android.com/apk/res/android"
    …>
    <TextView
        android:layout_width="wrap_content"
        android:layout_height="wrap_content"
        android:text="Hello World!" />
</RelativeLayout>
```

这样写虽然直观方便,但对于界面较复杂、规模较大的应用,其界面文本直接放在源文件代码中的各处,不利于应用程序的本地化,为此 Android Studio 建议的编程方式是:将一个应用界面上的全部文本都集中在 res\values 目录下的 strings.xml 文件中,作为"字符串类型的资源"统一管理和使用。

2. 配置字符串资源文件:strings.xml

我们对显示文本的 textView 控件和按钮 button 控件显示的内容均进行集中配置。

配置方法是:在当前工程 res\value 目录下,打开 strings.xml,在其中增加如下内容。

```
<resources>
    <string name="app_name">HelloWorld</string>
```

```xml
        <string name="action_settings">Settings</string>
        <string name="str_textView">Hello World!</string>
        <string name="str_button">开始课程</string>
</resources>
```

其中，<string/>标签的"name"属性所表示的是该字符串资源的引用（别名），在属性设置的"Resources"窗口中可以找到对应的工程。

3. 修改 textView 控件和 button 控件的"text"属性

经这样配置之后，把界面上的 textView 控件和 button 控件的"text"属性设置为所需文字对应字符串资源的引用即可，如图 2.50 所示。

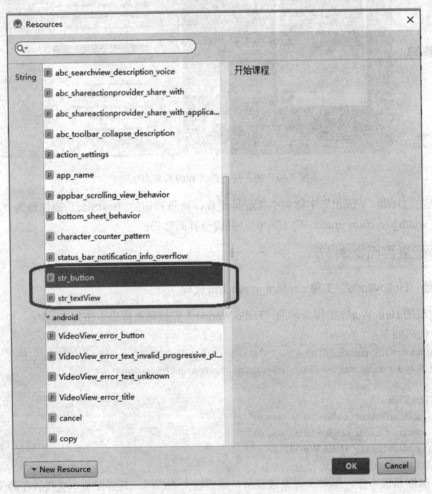

图 2.50　将控件的 text 属性设置为字符串资源的引用

此时可预览到程序界面上显示出引用的文字，如图 2.51 所示。

4. content_main.xml 文件内容

在可视化设计工程中界面所有变化均反映在 content_main.xml 文件中。

选择 res\layout 目录下的 content_main.xml，可点击 Editor 窗口底部的"Text"选项卡切换到代码视图查看，代码如下。

图 2.51　界面显示引用的字符串文本

```xml
<?xml version="1.0" encoding="utf-8"?>
<RelativeLayout xmlns:android="http://schemas.android.com/apk/res/android"
    ...>
<TextView
    android:layout_width="wrap_content"
    android:layout_height="wrap_content"
    android:text="@string/str_textView"                                //(3)
    android:textSize="@dimen/abc_text_size_headline_material"          //(2)
    android:textColor="@android:color/holo_red_dark"                   //(2)
    android:textStyle="bold"                                           //(2)
    android:id="@+id/textView" />                                      //(1)

    <ImageView
    android:layout_width="wrap_content"
    android:layout_height="300dp"                                      //(5)
    android:id="@+id/imageView"                                        //(1)
    android:src="@drawable/androidlover"                               //(4)
    android:layout_below="@+id/textView"
    android:layout_alignParentLeft="true"
    android:layout_alignParentStart="true" />

    <Button
        android:layout_width="wrap_content"
    android:layout_height="wrap_content"
    android:text="@string/str_button"                                  //(3)
    android:id="@+id/button"                                           //(1)
    android:textSize="@dimen/abc_action_button_min_width_overflow_material"  //(2)
    android:layout_below="@+id/imageView"
    android:layout_centerHorizontal="true" />
</RelativeLayout>
```

其中：

(1) android:id="@+id/textView":设置其 id 属性为"textView",代表该控件,以便在其后引用该控件。其他 2 个控件也有对应的 ID。
(2) 设置显示字符的大小、颜色和加粗。
(3) 显示内容引用了 strings.xml 中的文本资源对应字符串。
(4) 显示图片为工程 drawable 目录下的 androidlover 文件。
(5) 控制显示图片为 300 个像素。
其他属性在界面设计过程中为了控件对齐时生成的,后面会系统介绍。

2.2.3 代码编写与事件处理

设计好应用程序界面后,接下来就要为 Android 工程编写代码以实现其应用功能,同时应用还要求能响应用户的操作及各种外部事件。

Android Studio 开发的工程,其代码编写必须遵循一定的模式,在既有的开发框架中的特定位置添加代码。

1. 关联 UI 组件对象

Android 应用是基于 Java 语言的,已经设计好的界面控件,如果在代码中需要引用,需要首先定义成 Java 对象。通行的方法是声明成主程序类(位于 MainActivity.java 源文件中)的私有成员,并在初始化时将它们各自关联到界面上对应的控件。

为此需要改写 MainActivity.java 的代码如下(加黑语句为修改的内容)。

MainActivity.java:

```java
package com.easybooks.android.helloworld;
……                                              //系统生成的导入控件代码
import android.widget.Button;                     //导入命令按钮控件
import android.widget.ImageView;                  //导入图片显示控件
import android.widget.TextView;                   //导入文本显示控件

public class MainActivity extends AppCompatActivity {
    private TextView textView;                    //(1)定义显示"Hello World!"等文字的控件对象
    private ImageView imageView;                  //(1)定义显示"安卓小机器人"图片的控件对象
    private Button button_Start;                  //(1)定义"开始课程"按钮控件对象

    @Override
    protected void onCreate(Bundle savedInstanceState) {
        super.onCreate(savedInstanceState);
        setContentView(R.layout.activity_main);
        findViews();                              //(2)
        ……                                        //系统生成的其他代码
    }

    private void findViews() {                    //(3)
        textView = (TextView) findViewById(R.id.textView);
        imageView = (ImageView) findViewById(R.id.imageView);
        button_Start = (Button) findViewById(R.id.button);
    }
    ……                                            //系统生成的其他代码
}
```

其中:

（1）一个 Android 应用程序的 UI 组件（Palette 面板上）可分成 widget 与 layout 两大类，widget 组件是 UI 的最基本单位。换句话说，不能在这类组件内再放入其他的组件了，如本例用到的 TextView、ImageView 和按钮就都是 widget 组件。而 layout 组件是布局用组件，在它们的内部还可以放入其他组件。widget 组件相关的类都放在 android.widget 包内，当用户在编程过程中使用到这些组件时，系统自动生成"import android.widget.<组件类名>;"语句将相应的组件类引入程序，无须用户手工编写导入语句。

（2）每个 Android 应用都要在其主程序类中定义一个 findViews()方法，该方法的主要职能是将 UI 组件与 Java 对象关联起来，有时也会进行一些程序初始化的辅助工作。它一般在 onCreate()函数中被调用，且紧接在 setContentView()函数之后。

（3）findViews()方法内调用 findViewById()并指定 id 来获取界面源文件（content_main.xml）内相应的 UI 组件，然后赋值给对应的对象，之后这些对象就分别代表界面上的 UI 组件，可以在程序中引用了。

这里关联 UI 组件对象时所指定的 id 就是界面源文件（content_main.xml）中各 UI 元素的"android:id"属性值。

2. Android 事件处理

手机用户在大部分情况下会通过 UI 组件来输入，而这种互动可能会触发各种 UI 事件，所以在学习 Android 开发时，最重要的就是掌握 Android 程序的事件处理机制。事件处理机制是一种委托机制（delegation），就是将事件的监听与处理交给系统。Android 系统提供了两种处理事件的方式，现分别介绍如下。

（1）Java 传统机制

这是 Java 语言内置的事件处理机制，它的原理是：将要求响应的事件预先注册到系统监听器（俗称"看门狗"）中，同时编程实现该事件对应的处理方法。当程序运行时，系统监听器时刻处于监听状态，一旦监听到该事件发生，随即执行其对应的方法代码。而对于未注册的事件，即使它发生了，监听器也会对其一概忽略，不会有任何反应。

Android 用户最常用到的 UI 组件就是按钮（Button）组件，下面以 HelloWorld 工程中的按钮 button_Start 为例，说明 Java 传统事件处理的工作流程，如图 2.52 所示。

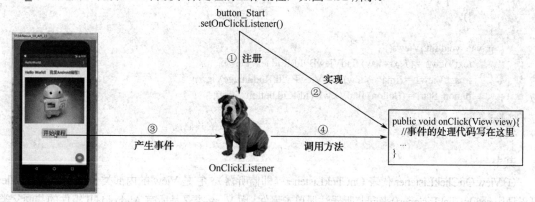

图 2.52　Java 传统事件处理流程

Java 传统事件处理流程如下。

① button_Start 向监听器注册事件：button_Start 必须先调用 setOnClickListener()向 OnClickListener（监听器）注册单击事件，这样在程序运行时，OnClickListener 才会去监听 button_Start 是否被单击。

② 实现单击事件的处理方法：事件处理方法用来实现用户单击 button_Start 后希望程序执行的功能，处理方法的代码写在 OnClickListener 的 onClick()方法内。

③ 产生单击事件：当界面上的 button_Start（"开始课程"按钮）被用户按下时，会产生一个"单击"事件，OnClickListener 监听器就会知道。

④ 执行处理：由于该单击事件事先已经向 OnClickListener 注册过了，故 OnClickListener 就会对它进行处理，于是自动调用用户编程实现好的 onClick()方法来响应用户按下按钮的操作。

根据以上原理，我们为 HelloWorld 工程添加单击事件处理功能，修改 MainActivity.java 的代码来实现上述事件处理机制如下（加黑语句为修改或添加的内容）。

MainActivity.java：

```java
……                                           //系统生成的其他代码
public class MainActivity extends AppCompatActivity {
    private TextView textView1;               //显示"Hello World!"等文字的控件
    private ImageView imageView1;             //显示"安卓小机器人"图片的控件
    private Button button_Start;              //"开始课程"按钮

    @Override
    protected void onCreate(Bundle savedInstanceState) {
        super.onCreate(savedInstanceState);
        setContentView(R.layout.activity_main);
        findViews();
        ……                                    //系统生成的其他代码
        fab.setOnClickListener(new View.OnClickListener() {
        ……                                    //系统生成的其他代码
        });
        button_Start.setOnClickListener(new View.OnClickListener() {  //①
            @Override
            public void onClick(View view) {                           //②
                String msg = textView.getText().toString().trim();     //③
                msg += "  我爱Android编程！";                            //④
                textView.setText(msg);                                 //⑤
            }
        });
    }
    private void findViews() {
        textView1 = (TextView) findViewById(R.id.textView);
        imageView1 = (ImageView) findViewById(R.id.imageView);
        button_Start = (Button) findViewById(R.id.button_Start);
    }
    ……                                        //系统生成的其他代码
}
```

其中：

① View.OnClickListener 代表 OnClickListener（即监听器），它是 View 的内部类（inner class）。View 可以调用 setOnClickListener()来向监听器注册单击事件，而 View 类又是所有 Android UI 组件的共同父类，故所有 UI 组件都可以这么做，因此这里 button_Start 才能顺利地调用 setOnClickListener()方法来注册事件。

② 利用匿名内部类实现 OnClickListener 的 onClick()方法，当按钮被按下时，onClick()方法会自动被调用并执行。onClick()方法必须传入一个 View 类型的参数 view，它代表的是触发事件的 UI 组件，这里就是被按下的 button_Start 按钮。

③ 通过 TextView 组件的 getText()方法获取其上的文字后，调用 trim()去除不必要的空格符号。

④ 定义字符串类型的变量 msg，串接上" 我爱Android编程！"在原"Hello World!"之后。

⑤ 通过调用 TextView 组件的 setText()方法将新组合的文字显示在界面上。

运行程序，结果如图 2.53（a）所示。单击"开始课程"，显示如图 2.53（b）所示。

(a)　　　　　　　　　　　　　(b)

图 2.53　传统机制实现单击事件处理

(2) Android 简易机制

在用上述的 Java 传统事件处理机制编程时，要求用户对 Java 的事件模型、工作原理及流程有一个比较清晰的认识，否则编写的程序极易出错。不过，Android 还提供了另一种更简易的事件处理方式，就是开发者可以在 layout 文件内设置 Button 组件被单击时要调用的方法，该方法可以自定义，效果等同于前面所说的 onClick()。如此一来，就省略了向监听器注册的步骤，简化了编程。

下面改用这种简单的方式实现事件处理，对 MainActivity.java 进行修改。

① 注释掉之前注册事件的代码段，在主程序类中添加一个自定义的 onStartClick()方法如下（加黑处为添加的方法）。

MainActivity.java：

```java
package com.easybooks.android.helloworld;
…
public class MainActivity extends AppCompatActivity {
    private TextView textView1;              //显示"Hello World!"等文字的控件
    private ImageView imageView1;            //显示"安卓小机器人"图片的控件
    private Button button_Start;             //"开始课程"按钮

    @Override
    protected void onCreate(Bundle savedInstanceState) {
        super.onCreate(savedInstanceState);
        setContentView(R.layout.activity_main);
        findViews();
        ……                                  //系统生成的其他代码
        //注释掉原来的代码段
        /*
        button_Start.setOnClickListener(new View.OnClickListener() {
            @Override
```

```
            public void onClick(View view) {
                String msg = textView.getText().toString().trim();
                msg += "    我爱 Android 编程！";
                textView.setText(msg);
            }
        });
        */
    }

    private void findViews() {
        ……
    }
    //添加的方法
    public void onStartClick(View view) {
        String msg = "简易型事件处理真方便！";
        textView.setText(msg);
    }
    ……                          //系统生成的其他代码
}
```

注意，用户自己添加定义的方法也必须传入一个 View 类型的参数，且该方法必须为"public void"类型。

② 切换到设计视图，选中手机屏上的"开始课程"按钮，在右边"Properties"子窗口中设置其"onClick"属性，选择该属性，可以看到其下拉列表选项中多了一条"onStartClick (…)"项——正是刚刚添加定义的方法，如图 2.54 所示，选中该项将其设为按钮的"onClick"属性值。

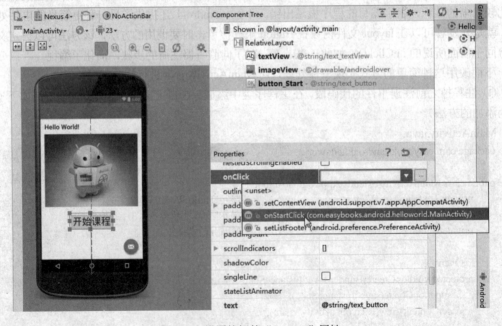

图 2.54　设置按钮的"onClick"属性

这样设置之后，打开设计界面的源文件 content_main.xml，可以看到按钮 button_Start 的"android:onClick"属性已经与用户定义的事件处理方法 onStartClick 相关联了，如下（加黑语句）所示。

content_main.xml：

```
<?xml version="1.0" encoding="utf-8"?>
<RelativeLayout xmlns:android="http://schemas.android.com/apk/res/android"
    …>
    <TextView
        …        />
    <ImageView
        …        />
    <Button
        
        android:onClick="onStartClick" />
</RelativeLayout>
```

最后运行程序，结果如图 2.55 所示。

（3）两种事件处理方式的比较

以上介绍的两种事件处理方式各有利弊，Java 传统机制直接将事件模型接口暴露给程序员，编程有一定的难度，但由于它所依赖的内部类 OnClickListener 是内置于所有 UI 组件的共同父类 View 中的，故一旦熟练掌握此方法，开发者就能灵活地运用它来实现 Android 中各种 UI 组件任何类型的事件处理功能。Android 简易机制则对用户屏蔽了 Java 内部的事件模型，使用户能够像 VB/VC#等桌面高级开发语言那样直接编写事件代码实现相应的功能，极大地简化和降低了编程难度。但这种方式有其局限性，仅适用于单击事件，而不像 Java 传统机制那样可以用于任何 UI 事件的处理。

在实际开发中，究竟采用哪一种方式处理事件，要视用户的自身情况（编程习惯、能力等）和工程需要而定。一般说来，如果工程规模较大，其中需要处理的事件类型又较为单一（多为按钮的单击事件），用简易机制比较划算。倘若对应用的交互要求很高，所涉及的 UI 组件种类、事件的类型十分多样，还是用 Java 传统机制比较好。

图 2.55　简易机制实现单击事件处理

3．本书实例代码规范

通过本章的学习，读者应该已经对 Android 开发的基本概念、Android 工程结构及编程方法有了一个全面的了解，在后续章节的讲解中，为了突出重点和避免赘述，对于所举实例的源程序，我们将只给出需要用户自己编写和添加的代码，而省略工程框架的代码，具体规范如下。

本书代码罗列规则：

- **界面源文件**：仅给出 content_main.xml 中各 UI 组件元素标签及其内容，不包括文件头"<?xml version="1.0" encoding="utf-8"?>"和排版布局框架"<RelativeLayout…></RelativeLayout>"标签的内容。
- **Java 源程序**：仅给出 MainActivity.java 中关联 UI 组件的私有成员对象定义、findViews()方法的代码，一般情况不包括主程序类框架中的其他代码，仅在这些代码有需要改动的时候，才给出完整的主程序类框架并将改动处加黑醒目地标示出来。
- **事件处理代码**：

（1）Java 传统机制

仅给出向监听器注册事件的完整代码段，形如：

```
<组件名>.setOnClickListener(new View.OnClickListener() {
    @Override
    public void onClick(View view) {
```

```
        //事件处理代码
        ……
    }
});
```
（2）Android 简易机制
仅给出用户添加自定义的事件处理方法的代码，形如：
```
public void <方法名>(View view) {
    //事件处理代码
    ……
}
```

2.3 升级 Android Studio 工程

因为在安装 Android Studio 的过程中总是通过连接网络安装最新的 Android SDK 组件，这样如果在该开发环境下打开此前的 Android Studio 工程，就不能直接编译运行。具体表现就是 ▶ 等按钮图标不可用，同时系统显示下列信息，例如在打开一个以前我们在另外一台计算机上创建的"RegisterPage"工程时显示如图 2.56 所示。

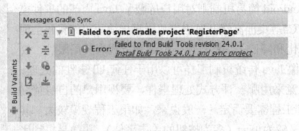

图 2.56　打开老的"RegisterPage"工程时

这时应该单击"Install Build Tools … and sync project"，系统显示接受许可条款如图 2.57 所示。

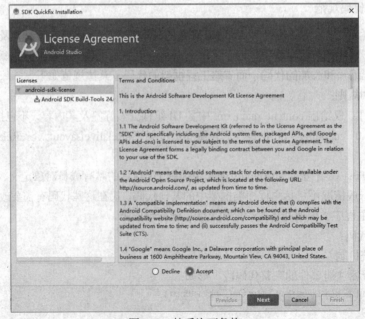

图 2.57　接受许可条款

用户选择"Accept",单击"Next",系统开始进行 Android Studio 组件安装。安装完成显示如图 2.58 所示。

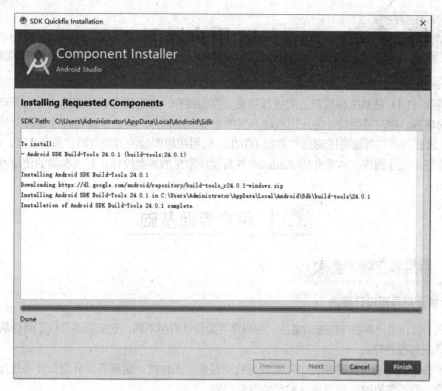

图 2.58　配套组件安装完成

这样就可以在当前环境下编译运行了。

第3章 Android 用户界面

用户界面（UI）是系统和用户之间进行信息交换的媒介，实现信息的内部形式与人类可以接受的形式之间的转换。用户界面设计是应用程序开发的重要组成部分，决定了应用程序是否美观、易用。

目前，流行的用户界面是图形用户界面（GUI），采用图形图像的方式与用户进行交互，通过窗口、菜单和按钮等来进行操作。本章介绍 Android 界面设计常见的界面控件、界面布局的使用方法及其初步功能代码。

3.1 用户界面基础

3.1.1 用户界面基本要求

1. 手机应用界面设计要求

在手机上设计用户界面与传统的桌面应用程序界面设计有所不同，它必须满足以下两点基本要求。

（1）界面与程序分离

现实中的手机界面设计者和程序开发者是独立且并行工作的，这就需要界面设计与程序逻辑完全分离，修改界面不需要改动程序功能实现的逻辑代码。

（2）自适应手机屏幕

不同型号手机的屏幕可视参数（如解析度、尺寸和长宽比等）各不相同，故手机应用也要能根据不同的屏幕参数，自动调整（自适应）其界面控件的位置和尺寸，构造出符合人机交互规则的用户界面，避免出现凌乱、拥挤的情况。

2. Android 用户界面支持

Android 系统为手机用户界面的开发提供了强有力的支持。在界面设计与程序逻辑分离方面，Android 使用 XML 文件对用户界面进行描述，各种资源文件分门别类地独立保存于各自专有的文件夹中。例如，描述用户界面的源文件 content_main.xml 存放在项目 res\layout 文件夹下，而实现程序逻辑的 Java 源文件 MainActivity.java 则放在 java 目录中，两者是完全分离的。在界面自适应方面，Android 允许模糊定义界面控件的位置和尺寸，通过声明界面控件的相对位置和粗略尺寸，使界面控件能够根据屏幕尺寸和屏幕摆放方式动态地调整显示方式。

3. Android 用户界面框架

（1）MVC 模型

Android 用户界面框架采用 MVC（Model-View-Controller）模型，为用户界面提供了处理用户输入的控制器（Controller）和显示图像的视图（View），模型（Model）是应用程序的核心，数据和代码被保存在模型中。控制器、视图和模型的关系如图 3.1 所示。

- MVC 模型中的视图

MVC 模型中的视图将应用程序的信息反馈给用户，可能的反馈方式包括视觉、听觉或触觉等，但最常用的就是通过屏幕显示反馈信息。Android 系统的界面控件以一种树形结构组织在一起，这种树形结构称为视图树（View Tree），如图 3.2 所示。

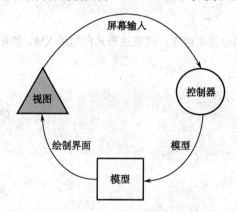

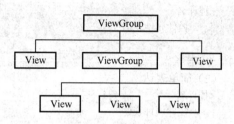

图 3.1　Android 用户界面框架的 MVC 模型　　　　　图 3.2　视图树

视图树由 View 和 ViewGroup 构成。View 是界面中最基本的可视单元，存储了屏幕上特定矩形区域内所显示内容的数据结构，并能够实现所占据区域的界面绘制、焦点变化、用户输入和界面事件处理等功能。View 也是一个重要的基类，所有在界面上的可见控件都是 View 的子类。ViewGroup 是一种能够承载多个 View 的显示单元，一般有两个用途：一个是承载界面布局，另一个是承载具有原子特性的重构模块。

Android 系统在屏幕上绘制界面控件时，会依据视图树的结构从上至下绘制每一个界面控件。每个控件负责对自身的绘制，如果控件包含子控件，该控件会通知其下所有子控件进行绘制。

● MVC 模型中的控制器

MVC 模型中的控制器能够接收并响应程序的外部动作，如按键动作或触摸屏幕动作等。控制器使用队列处理外部动作，每个外部动作作为一个独立的事件被加入队列中，然后 Android 用户界面框架按照"先进先出"的规则从队列中获取事件，并将这个事件分配给所对应的事件处理函数。

（2）单线程用户界面

单线程用户界面是 Android 用户界面框架的另一个重要概念。在单线程用户界面中，控制器从队列中获取事件，视图在屏幕上绘制用户界面，使用的都是同一个线程。

● 优点

单线程用户界面能够降低应用程序的复杂程度，同时也能降低开发的难度。因为用户不需要在控制器和视图之间进行同步，同时所有事件处理完全按照其加入队列的顺序进行。也就是说，在事件处理函数返回前不会处理其他事件，因此用户界面的事件处理函数具有原子性。

● 缺点

在单线程用户界面下，如果事件处理函数过于复杂，可能会导致用户界面失去响应。因此在实际编程中应当尽可能在事件处理函数中使用简短的代码，或将复杂的工作交给后台线程处理。

3.1.2　控件概述

Android 系统的界面控件分为定制控件和系统控件。定制控件是用户独立开发或通过继承并修改系统控件后所产生的新控件，能够提供特殊的功能和显示需求。系统控件则是 Android 系统中已经封装好的界面控件，是应用程序开发过程中最常见的功能控件。系统控件更有利于进行快速开发，同时能够使 Android 应用程序的界面保持一定的一致性。

常见的系统控件包括 TextView、EditText、Button、ImageButton、CheckBox、RadioButton、Spinner 和 ListView 等。

1. 控件的表达

在 Android Studio 中,用 XML 文件中的 UI 控件来描述界面控件,其表达形式自然是 XML 的标签,有如下两种写法。

(1) 规范表达

```
<控件名
    控件属性表达
    ……>
</控件名>
```

(2) 简单表达

```
<控件名
    控件属性表达
    ……/>
```

这两种形式是等价的,用户可视编程和代码结构的可读性需要来灵活地使用这两种写法。

2. 控件的属性

控件的属性直接在其控件标签内赋值,格式如下。

```
<控件名
    android:id="@+id/<名称>"
    android:<属性名 1>="<值 1>"
    android:<属性名 2>="<值 2>"
    ……
    android:<属性名 n>="<值 n>"
</控件名>
```

每个控件可供赋值设置的属性个数和类型都不同,这里列举几种最为通用的属性,部分属性前面已经用过多次,这里进行系统说明。

(1) 指定控件标识:android:id 属性。

android:id="@+id/TextView1":表示新建立 ID 资源 TextView1。

android:id="@android: id/TextView1":表示不是新添加的资源,或属于 Android 框架的资源,必须添加 Android 包的命名空间。

指定控件 id 属性是为了在其后引用该控件,当前界面的所有控件 id 属性的值不能相同。

(2) 指定控件的大小:android:layout_width 属性设置宽度;android:layout_height 属性设置高度。

android:layout_width="wrap_content":只要能够包含所显示的字符串即可。

android:layout_width="fill_content":等于父控件的宽度。

android:layout_height="240dp":240 个像素。

(3) 指定控件默认显示的字符串:android:text 属性。在运行过程中控件显示的字符串可以通过 setText()函数去修改。

android:text="abc 汉字":显示"abc 汉字"。

android:text="@string/text1":显示字符串资源文件(strings.xml)中的"text1"标识符指定的内容。

(4) 指定控件显示字符的大小:android:textSize 属性:例如,android:textSize="@dimen/abc_text_size_display_1_material"。一串符号对应的字符的大小通过不断试验才能知道。

(5) 指定控件位置:例如,android:layout_below="@+id/EditText1":该控件位于 EditText1 控件的

下部。同样，还有指定位于控件上部、左部和右部的属性。

（6）指定控件对齐位置：例如，android:layout_alignLeft="@+id/Button1"：该控件与 Button1 控件左对齐。同样，还有与指定控件上对齐、下对齐和右对齐的属性。

还有，android:layout_alignParentLeft="true"：将该控件的左部与其父控件的左部对齐。另外还有属性设置将该控件的右部、上部、下部与其父控件相应的部位对齐的属性。

3. 控件的事件

控件除了单击事件（ClickEvent），还有按键事件（KeyEvent）、触摸事件（TouchEvent）等。在这些事件发生时，Android 用户界面框架调用界面控件的事件处理函数对事件进行处理。在 MVC 模型中，控制器根据事件类型不同，将其传递给控件不同的事件处理函数，例如，单击事件传递给 onClick()函数；按键事件传递给 onKey()函数；触摸事件传递给 onTouch()函数。

3.2 基本的界面控件

3.2.1 字符显示和编辑控件：TextView/EditText

字符显示和编辑控件统称文本框，TextView 控件用于显示字符；EditText 控件继承于 TextView，但用来输入和编辑字符。

【例 3.1】设计一个字符显示和编辑例子，第一行显示"用户名："信息，第二行输入用户名。效果如下：

用户名：

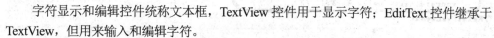

1. 定义文本框

创建一个默认为"Basic Activity" APP 工程，系统生成的 MainActivity.java 文件和主设计界面文件 content_main.xml，其中 content_main.xml 内容如图 3.3 所示。

按照要求设计界面，或者直接输入下列代码取代红框中的代码：

```
<TextView android:id="@+id/textView"
    android:layout_width="wrap_content"
    android:layout_height="wrap_content"
    android:text="用户名："
    android:textSize="@dimen/abc_text_size_display_1_material">
</TextView>
<EditText android:id="@+id/editText"
    android:layout_width="fill_parent"
    android:layout_height="wrap_content"
    android:text="                "
    android:layout_below="@+id/textView"
    android:textSize="@dimen/abc_text_size_display_1_material">
</EditText>
```

2. 引用文本框

在 MainActivity.java 文件中的 findViews()方法中引用 XML 文件中定义的 textView 和 editText，并更改其默认的显示内容。

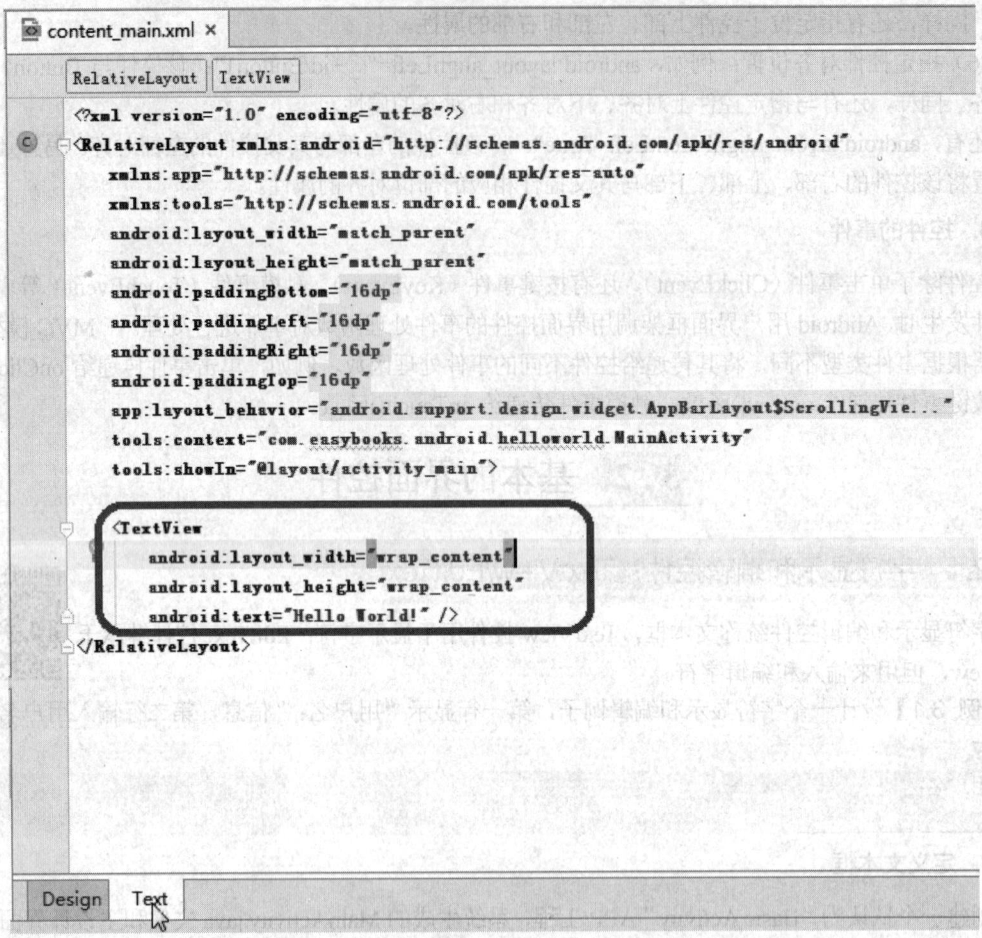
content_main.xml ×
RelativeLayout
TextView
<?xml version="1.0" encoding="utf-8"?>
<RelativeLayout xmlns:android="http://schemas.android.com/apk/res/android"
xmlns:app="http://schemas.android.com/apk/res-auto"
xmlns:tools="http://schemas.android.com/tools"
android:layout_width="match_parent"
android:layout_height="match_parent"
android:paddingBottom="16dp"
android:paddingLeft="16dp"
android:paddingRight="16dp"
android:paddingTop="16dp"
app:layout_behavior="android.support.design.widget.AppBarLayout$ScrollingVie..."
tools:context="com.easybooks.android.helloworld.MainActivity"
tools:showIn="@layout/activity_main">
<TextView
android:layout_width="wrap_content"
android:layout_height="wrap_content"
android:text="Hello World!" />
</RelativeLayout>
Design
Text

图 3.3　主设计界面文件中的控件

```
package com.easybooks.android.helloworld;
import …                                                          //(a)
public class MainActivity extends AppCompatActivity {
    //定义控件变量
    private TextView myTextView;                                  //(a)
    private EditText myEditText;                                  //(a)

    @Override
    protected void onCreate(Bundle savedInstanceState) {
        super.onCreate(savedInstanceState);
        setContentView(R.layout.activity_main);
        findViews();              //调用 findViews()函数查找界面控件对象
        ……                        //系统生成的其他代码
}

    private void findViews() {
        myTextView = (TextView) findViewById(R.id.textView);      //(b)
        myEditText = (EditText) findViewById(R.id.editText);      //(b)
        myTextView.setText("用户名：");                            //(c)
        myEditText.setText("ZhouHeJun");                          //(c)
```

```
    }
        ……                                           //系统生成的其他代码
    }
```

其中：

（a）输入定义控件变量的语句，系统在 import 处会自动增加相应的控件类包。如果从其他文件中复制定义控件变量内容粘贴到此处，需要用户一个一个按 Alt+Enter 才能分别导入。

（b）findViewById()函数：通过在 XML 文件中定义的控件的 ID（textView 和 editText）引用赋值给相应的控件变量 myTextView 和 myEditText 中。

（c）setText()函数：用来修改 myTextView 和 myEditText 对应的控件对象所显示的内容。

3．运行效果

运行程序，TextView 控件和 EditText 控件的显示效果如图 3.4 所示。

图 3.4　TextView 和 EditText 控件

3.2.2　按钮和图像按钮控件：Button/ImageButton

Button 是按钮控件，用户在该控件上点击，引发相应的事件处理函数。通常 Button 按钮控件上显示字符，如果想要在按钮控件上显示图像，可使用 ImageButton 控件。

【例 3.2】设计按钮和图像按钮例，初始界面如图 3.5（a）所示，单击"OK"按钮，显示如图 3.5（b）所示。单击打钩图标，显示如图 3.5（c）所示。

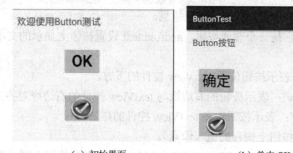  

　　　　（a）初始界面　　　　　　　　　（b）单击 OK　　　　　　　　　（c）单击打钩

图 3.5　按钮

1．定义按钮

按照要求设计界面，或者直接在 XML 文件（content_main.xml）中输入如下代码。

```
<TextView android:id="@+id/textView"
    android:layout_width="wrap_content"
    android:layout_height="wrap_content"
    android:text="@string/hello"                                        //(a)
    android:textSize="@dimen/abc_text_size_large_material">
</TextView>
<Button android:id="@+id/button"                                       //(b)
    android:layout_width="wrap_content"
    android:layout_height="wrap_content"
    android:text="OK"
    android:textSize="@dimen/abc_action_button_min_width_overflow_material"
    android:layout_below="@+id/textView"
    android:layout_alignRight="@+id/textView"
```

```
        android:layout_alignEnd="@+id/textView"
        android:layout_marginTop="45dp">
</Button>
<ImageButton android:id="@+id/imageButton"                                          //(c)
        android:layout_width="wrap_content"
        android:layout_height="wrap_content"
        android:src="@drawable/yes"
        android:layout_marginTop="41dp"
        android:layout_below="@+id/button"
        android:layout_alignRight="@+id/button"
        android:layout_alignEnd="@+id/button"
        android:layout_alignLeft="@+id/button"
        android:layout_alignStart="@+id/button">
</ImageButton>
```

其中：

（a）ID 为 textView 字符显示控件，因为初始要显示的标题文字定义：android:text="@string/hello"，所以显示的内容定义在 strings.xml 文件中 name="hello"属性<string … ></string>标签中的内容。在 res\values 目录打开 strings.xml 文件，加入如下内容。

```
<resources>
    ……
    <string name="hello">欢迎使用 Button 测试</string>
</resources>
```

在程序运行过程中，textView 控件用于显示用户点击的按钮信息。

（b）定义 ID 为 button 的 Button 控件，它是一个普通按钮，android:text 设置按钮上显示的文字为"OK"。其他属性如下。

android:layout_below="@+id/textView"：表示按钮位于 textView 控件的下方。

android:layout_alignRight="@+id/textView"：表示按钮的右边缘与 textView 控件的右边缘对齐。

android:layout_alignEnd="@+id/textView"：表示按钮对齐 textView 控件的尾部。

android:layout_marginTop="45dp"：指定按钮上偏移的值（像素）。

（c）定义了 ImageButton 控件（图像按钮），android:src="@drawable/yes"：指定按钮上图片来源 drawable 目录，图片文件名为 yes.png，扩展名可以是 jpg、gif、png、bmp 等。

在程序代码中，需要更换为按钮图片 crystal.jpg，必须预先将 yes.png 和 crystal.jpg 图片文件拷贝到当前工程的 res\drawable 目录中。

其他属性如下：

android:layout_alignLeft="@+id/button"：按钮的左边缘与 button 按钮的左边缘对齐。

android:layout_alignStart="@+id/button"：按钮对齐 button 按钮的头部。

特别注意：如果 content_main.xml 文件中有显示红色的部分，表示系统没有找到有关内容。例如，android:text="@string/hello" 没有在 strings.xml 文件中找到 hello 对应的字符串；android:src="@drawable/yes"没有在 drawable 中找到 yes 文件。

2. 引用和更改按钮

在 MainActivity.java 文件中分别编写下列代码。

（1）定义控件变量

```
//定义控件变量
private TextView myTextView;
```

```
private Button myButton;
private ImageButton myImageButton;
```

（2）在 findViews()方法中引用 XML 文件中定义的控件变量。

```
private void findViews() {
myTextView = (TextView) findViewById(R.id.textView);
myButton = (Button) findViewById(R.id.button);
myImageButton = (ImageButton) findViewById(R.id.imageButton);
}
```

在 onCreate()方法中加入调用 findViews()方法。

（3）要更改按钮上的文字和图片，可以采用两种不同的事件处理机制，除了可以在按钮的单击事件处理过程中外，也可以将多个按钮注册到同一个单击事件的监听器上。

在 onCreate()函数中编写代码如下。

```
Button.OnClickListener bListener = new Button.OnClickListener() {    //(a)
    @Override
    public void onClick(View view) {
        switch (view.getId()) {
            case R.id.button:                                         //(b)
                myTextView.setText("Button 按钮");
                myButton.setText("确定");
                myImageButton.setImageResource(R.drawable.yes);
                return;
            case R.id.imageButton:                                    //(c)
                myTextView.setText("ImageButton 按钮");
                myButton.setText("OK");
                myImageButton.setImageResource(R.drawable.crystal);
                return;
        }
    }
};
myButton.setOnClickListener(bListener);                               //(d)
myImageButton.setOnClickListener(bListener);                          //(e)
```

其中：

（a）定义了一个名为 bListener 的单击事件监听器。

（b）myTextView 显示"Button 按钮"；将 myButton 引用的按钮的显示内容更改为"确定"；将 yes.png 文件传递给 myImageButton 引用的图像按钮。

（c）myTextView 显示"ImageButton 按钮"；将 myButton 引用的按钮的显示内容更改为"OK"；将 crystal.jpg 文件传递给 myImageButton 引用的图像按钮。

（d）将监听器注册到 myButton 上。

（e）将监听器注册到 myImageButton 上。

3．运行效果

运行程序，屏幕上显示 Button 按钮和 ImageButton 图像按钮。点击 Button 按钮，其上文字由"OK"变为"确定"；点击 ImageButton 按钮，按钮上的图片更换为一个圆形水晶图案；同时屏幕上方的字符显示控件会实时地显示当前用户点击了哪一个按钮。

3.2.3 复选框：CheckBox

CheckBox 控件称为复选框，它有两种状态：在非选择状态，控件显示为一个方框；而在选择状态，控件显示方框中包含打钩。界面上可以同时放置多个 CheckBox 控件，每个 CheckBox 控件的状态之间互相没有影响，一般用于表示可以选择多个选项的栏目。

【例 3.3】设计复选框例，初始界面如图 3.6（a）所示；单击"南京"和"苏州"复选框，显示如图 3.6（b）所示。

图 3.6 复选框

1. 定义复选框

按照要求设计界面，或者直接在 XML 文件（content_main.xml）中输入如下代码。

```
<TextView android:id="@+id/textView"
    android:layout_width="wrap_content"
    android:layout_height="wrap_content"
    android:text="江苏省 GDP 排名前三的城市有："
    android:textSize="@dimen/abc_text_size_large_material">
</TextView>
<CheckBox
    android:layout_width="wrap_content"
    android:layout_height="wrap_content"
    android:text="南京"
    android:id="@+id/checkBox1"
    android:layout_below="@+id/textView"
    android:layout_alignParentLeft="true"
    android:layout_alignParentStart="true"
    android:textSize="@dimen/abc_text_size_large_material" />
<CheckBox
    android:layout_width="wrap_content"
    android:layout_height="wrap_content"
    android:text="苏州"
    android:id="@+id/checkBox2"
    android:layout_below="@+id/checkBox1"
    android:layout_alignParentLeft="true"
    android:layout_alignParentStart="true"
    android:textSize="@dimen/abc_text_size_large_material" />
<CheckBox
    android:layout_width="wrap_content"
    android:layout_height="wrap_content"
    android:text="无锡"
```

```
android:id="@+id/checkBox3"
android:layout_below="@+id/checkBox2"
android:layout_alignParentLeft="true"
android:layout_alignParentStart="true"
android:textSize="@dimen/abc_text_size_large_material" />
```

这里定义了 3 个复选框，ID 分别为 checkBox1、checkBox2 和 checkBox3，提供"南京""苏州""无锡"三个城市让用户选择，可以多选。

2. 使用复选框

（1）在 MainActivity.java 中定义控件变量

```
private TextView myTextView;
private CheckBox myCheckBox1;
private CheckBox myCheckBox2;
private CheckBox myCheckBox3;
private String msg;
```

（2）在 findViews()方法中引用 XML 文件中定义的 TextView 及三个 CheckBox。

代码如下：

```
private void findViews() {
    myTextView = (TextView) findViewById(R.id.textView);
    myCheckBox1 = (CheckBox) findViewById(R.id.checkBox1);      //""南京"复选框
    myCheckBox2 = (CheckBox) findViewById(R.id.checkBox2);      //""苏州"复选框
    myCheckBox3 = (CheckBox) findViewById(R.id.checkBox3);      //""无锡"复选框
    msg = myTextView.getText().toString().trim() + "\n";
}
```

其中，"msg = myTextView.getText().toString().trim() + "\n";"获取 myTextView 的文本，msg 是主程序类的成员变量，根据用户所选复选框的不同，存储动态变化的选项信息，它与各控件对象在一起声明。

在 onCreate()方法中加入调用 findViews()方法。

（3）CheckBox 设置单击（勾选）事件监听器的方法为 CheckBox.OnClickListener。根据监听到的用户选择哪一个复选框，在文本框中显示加入的项目信息。

代码如下：

```
CheckBox.OnClickListener cListener = new CheckBox.OnClickListener() {
    @Override
    public void onClick(View view) {
        switch (view.getId()) {
            case R.id.checkBox1:
                if (myCheckBox1.isChecked()) {                          //(a)
                    msg += myCheckBox1.getText().toString() + " ";      //(b)
                    myTextView.setText(msg);                            //(c)
                }
                return;
            case R.id.checkBox2:
                if (myCheckBox2.isChecked()) {
                    msg += myCheckBox2.getText().toString() + " ";
                    myTextView.setText(msg);
                }
                return;
            case R.id.checkBox3:
```

```
                if (myCheckBox3.isChecked()) {
                    msg += myCheckBox3.getText().toString() + " ";
                    myTextView.setText(msg);
                }
                return;
            }
        }
    };
    myCheckBox1.setOnClickListener(cListener);
    myCheckBox2.setOnClickListener(cListener);
    myCheckBox3.setOnClickListener(cListener);                                    //(d)
```

可以看到，在 switch 语句中每个 case 块对复选框被点击事件的处理过程基本都是一样的，其中：
（a）isChecked()方法判断该复选框是否被选中。
（b）将被选中选项的文本（城市名）添加到 msg 字符串的末尾。
（c）在 myTextView 对应的文本框中显示用户选择的城市（可以多个）。
（d）将监听器注册到各 CheckBox 上。

3. 运行效果

运行程序，屏幕上显示三个城市的 CheckBox 选项。点击勾选城市名，被选中的城市名称依次添加到上方的字符显示控件中。

3.2.4 单选按钮及其容器：RadioButton 和 RadioGroup

RadioButton 控件也有两种状态：在非选择状态，控件显示为一个空心小圆圈；在选择状态，该小圆圈中会包含一个点。一般界面上可以同时放置多个 RadioButton 控件组成单选按钮组。

RadioGroup 控件是 RadioButton 控件的承载体（称为容器）。在一个 RadioGroup 中，用户仅能够选择其中的一个 RadioButton。RadioGroup 控件在程序运行时看不见。应用程序中可能包含一个或多个 RadioGroup。RadioGroup 位于工具箱面板的"Containers"（容器）条目下。

【例 3.4】设计单选按钮及其容器例，选的是"江苏省 GDP 排名第一的城市"，只有一个，提供"南京""苏州""无锡"三个城市让用户选择。

初始界面如图 3.7（a）所示。单击"苏州"单选按钮，显示如图 3.7（b）所示。

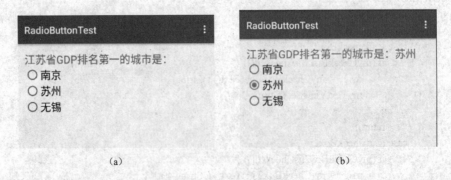

图 3.7 单选按钮及其容器组合

一个 RadioGroup 中包含三个 RadioButton，它们是互斥的，即程序运行时，选择其中任一个，其余两个都会自动变为非选择状态。

1. 定义单选按钮

在 XML 文件（content_main.xml）中，编写代码：

```xml
<TextView android:id="@+id/textView"
    android:layout_width="wrap_content"
    android:layout_height="wrap_content"
    android:text="江苏省 GDP 排名第一的城市是："
    android:textSize="@dimen/abc_text_size_large_material">
</TextView>
<RadioGroup
    android:layout_width="match_parent"
    android:layout_height="match_parent"
    android:layout_below="@+id/textView"
    android:layout_alignParentLeft="true"
    android:layout_alignParentStart="true">
    <RadioButton
        android:layout_width="wrap_content"
        android:layout_height="wrap_content"
        android:text="南京"
        android:id="@+id/radioButton1"
        android:textSize="@dimen/abc_text_size_large_material" />
    <RadioButton
        android:layout_width="wrap_content"
        android:layout_height="wrap_content"
        android:text="苏州"
        android:id="@+id/radioButton2"
        android:textSize="@dimen/abc_text_size_large_material" />
    <RadioButton
        android:layout_width="wrap_content"
        android:layout_height="wrap_content"
        android:text="无锡"
        android:id="@+id/radioButton3"
        android:textSize="@dimen/abc_text_size_large_material" />
</RadioGroup>
```

其中：<RadioGroup>标签声明了一个 RadioGroup，里面定义了 3 个 RadioButton，ID 分别为 radioButton1、radioButton2 和 radioButton3，它们是 RadioGroup 的子控件。

2. 引用单选按钮

（1）在 MainActivity.java 文件中的定义

```java
private TextView myTextView;
    private RadioButton myRadioButton1;
    private RadioButton myRadioButton2;
    private RadioButton myRadioButton3;
    private String msg;
```

（2）findViews()方法中引用 XML 文件中定义的 TextView 和 3 个 RadioButton。

代码如下：

```java
private void findViews() {
    myTextView = (TextView) findViewById(R.id.textView);
    myRadioButton1 = (RadioButton) findViewById(R.id.radioButton1); //"""南京"单选按钮
    myRadioButton2 = (RadioButton) findViewById(R.id.radioButton2); //"""苏州"单选按钮
```

```
            myRadioButton3 = (RadioButton) findViewById(R.id.radioButton3); //"""无锡"单选按钮
            msg = myTextView.getText().toString().trim();
    }
```

在 onCreate()方法中加入调用 findViews()方法。

（3）RadioButton 设置单击（点选）事件监听器的方法为 RadioButton.OnClickListener。根据监听到的用户选择哪一个单选按钮，在文本框中显示选择的项目信息。

代码如下：

```
RadioButton.OnClickListener rListener = new RadioButton.OnClickListener() {
    @Override
    public void onClick(View view) {
        switch (view.getId()) {
            case R.id.radioButton1:
                if (myRadioButton1.isChecked()) {                              //(a)
                    String text = msg + myRadioButton1.getText().toString();   //(b)
                    myTextView.setText(text);
                }
                return;
            case R.id.radioButton2:
                if (myRadioButton2.isChecked()) {
                    String text = msg + myRadioButton2.getText().toString();
                    myTextView.setText(text);
                }
                return;
            case R.id.radioButton3:
                if (myRadioButton3.isChecked()) {
                    String text = msg + myRadioButton3.getText().toString();
                    myTextView.setText(text);
                }
                return;
        }
    }
};
myRadioButton1.setOnClickListener(rListener);
myRadioButton2.setOnClickListener(rListener);
myRadioButton3.setOnClickListener(rListener);                                  //(c)
```

这里 switch 语句的 case 块对每个单选按钮被点选事件的处理过程也都是一样的，其中：

（a）isChecked()方法判断该单选按钮是否被选中。

（b）由于每次只能选中一个单选按钮，且只显示一个城市，故这里用一个局部变量 text 暂存当前选中的城市名，并将其与 msg 字符串拼接后在 TextView 中显示。

（c）将监听器注册到各 RadioButton 上。

3. 运行效果

运行程序，屏幕上显示三个城市的 RadioButton 选项。点选城市名，被选中的城市名称显示在字符显示控件中。

3.2.5 下拉列表：Spinner

Spinner 是从多个选项中选择一个选项的控件，类似于桌面程序的组合框（ComboBox），但没有组合框的下拉菜单，而是使用浮动菜单为用户提供选择。

【例 3.5】 在页面上设计下拉列表,包含"博士"、"硕士"和"学士"列表项。用户单击下拉列表,展开列表项;选择列表项,下拉列表显示选择的列表项。

1. 定义下拉列表

在 XML 文件(content_main.xml)中编写代码:

```
<Spinner
    android:layout_width="300dip"
    android:layout_height="wrap_content"
    android:id="@+id/spinner"
    android:layout_alignParentTop="true"
    android:layout_alignParentLeft="true"
    android:layout_alignParentStart="true" />
```

其中:<Spinner>标签声明了一个 Spinner 控件,并指定该控件的宽度为 300dip。

2. ArrayAdapter 适配器

(1)在 MainActivity.java 文件中定义控件变量。

```
private Spinner mySpinner;
```

(2)在 findViews()方法中,定义一个 ArrayAdapter 适配器,在 ArrayAdapter 中添加在 Spinner 中可以选择的内容。

代码如下:

```
private void findViews() {
    mySpinner = (Spinner) findViewById(R.id.spinner);
    List<String> list = new ArrayList<String>();                                    //(a)
    list.add("博士");                                                                //(b)
    list.add("硕士");                                                                //(b)
    list.add("学士");                                                                //(b)
    ArrayAdapter<String> adapter1 = new ArrayAdapter<String>(this, android.R.layout.simple_spinner_item, list);                                                                           //(c)
    adapter1.setDropDownViewResource(android.R.layout.simple_spinner_dropdown_item); //(d)
    mySpinner.setAdapter(adapter1);                                                 //(e)
}
```

其中:

(a)创建了一个字符串数组列表(ArrayList),这种数组列表可以根据需要进行增减,<String>表示数组列表中保存的是字符串类型的数据。

(b)使用 add()函数分别向数组列表中添加三个字符串。

(c)创建了一个 ArrayAdapter 类型的数组适配器,它将界面控件 Spinner 和底层数据 ArrayList 绑定在一起,这样所有 ArrayList 中的数据都将显示在 Spinner 的浮动菜单中。

(d)是 Android 系统内置的一种浮动菜单。在使用适配器绑定界面控件和底层数据后,应用程序就不需要再监视底层数据的变化,用户界面显示的内容与底层数据一致。

(e)实现适配器绑定过程。

在 onCreate()方法中加入调用 findViews()方法。

3. 运行效果

运行程序,屏幕上显示一个下拉列表,初始项为"博士",点右边的下箭头,出现浮动菜单,内有"博士"、"硕士"和"学士"三个选项,如图 3.8 所示。

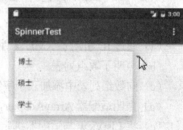

图 3.8 下拉列表

3.2.6 列表框：ListView

ListView 是用于垂直显示文本选项的列表控件。如果显示内容超过控件范围，会出现垂直滚动条。ListView 也是通过适配器将数据和显示控件绑定的，它能在有限的屏幕上提供大量内容供用户选择，而且支持点击事件，用少量的代码就可以实现复杂的选择功能。

【例 3.6】在页面上设计字符串显示框和列表框，列表框包含"博士"、"硕士"和"学士"列表项。用户单击列表项，字符串框显示选择的列表项。

1. 定义列表框

在 XML 文件（content_main.xml）中，编写代码：

```xml
<TextView android:id="@+id/textView"
    android:layout_width="fill_parent"
    android:layout_height="wrap_content"
    android:text="列表视图信息"
    android:textSize="@dimen/abc_text_size_medium_material" />
<ListView android:id="@+id/listView"
    android:layout_width="wrap_content"
    android:layout_height="wrap_content"
    android:layout_below="@+id/textView">
</ListView>
```

其中：<ListView>标签声明了一个 ListView 控件，ID 为 myListView。

2. ArrayAdapter 适配器

（1）在 MainActivity.java 文件中定义控件变量

```java
private TextView myTextView;
private ListView myListView;
```

（2）findViews()方法中为 ListView 创建适配器，并添加 ListView 中所显示的内容。

代码如下：

```java
private void findViews() {
    myTextView = (TextView) findViewById(R.id.textView);
    myListView = (ListView) findViewById(R.id.listView);                             //(a)
    List<String> list = new ArrayList<String>();                                      //(b)
    list.add("博士");                                                                  //(c)
    list.add("硕士");                                                                  //(c)
    list.add("学士");                                                                  //(c)
    ArrayAdapter<String> adapter1 = new ArrayAdapter<String>(this, android.R.layout.simple_list_item_1, list);
                                                                                      //(d)
    myListView.setAdapter(adapter1);                                                  //(e)
}
```

其中：

（a）通过 ID 引用了 XML 文件中声明的 ListView。

（b）声明了数组列表。

（c）向数组列表中添加三个字符串。

（d）声明适配器 ArrayAdapter，第 3 个参数 list 说明适配器的数据源为数组列表。

（e）将 ListView 与适配器绑定。

在 onCreate()方法中加入调用 findViews()方法。

3. 事件监听器

在 onCreate()方法中声明 ListView 子项的点击事件监听器,用以判断用户在 ListView 中选择的是哪一个子项,并将这些信息在文本框中显示。

代码如下:

```
AdapterView.OnItemClickListener vListener = new AdapterView.OnItemClickListener() {    //(a)
    @Override
    public void onItemClick(AdapterView<?> arg0, View arg1, int arg2, long arg3) {     //(b)
        String msg="" ";
        if(arg2==0)   msg="博士";                                                       //(c)
        if(arg2==1)   msg="""""""硕士";
        if(arg2==2)   msg="""""""学士";
        myTextView.setText(msg);                                                        //(d)
    }
};
myListView.setOnItemClickListener(vListener);                                           //(e)
```

其中:

(a) AdapterView.OnItemClickListener 是 ListView 子项的点击事件监听器,同样是一个接口,需要实现 onItemClick()函数。

(b) onItemClick()函数中一共有 4 个参数。

参数 1:适配器控件(ListView)。

参数 2:适配器内部的控件(ListView 中的子项)。

参数 3:适配器内部的控件(子项)的位置,从 0 开始。

参数 4:子项的行号,也从 0 开始。

(c) 把(子项父控件的信息、子控件信息、位置信息和 ID 信息)各参数内容转换为字符串并将它们连接起来。

(d) 把连接起来的字符串显示在 TextView 控件文本框中。

(e) ListView 指定声明的监听器。

4. 运行效果

运行程序,屏幕上显示一个列表,内容为 "博士" "硕士" "学士",点击其中任一项,上方文本框中显示该项所对应的父控件、子控件、位置和 ID 信息这 4 个参数的内容,如图 3.9 所示。

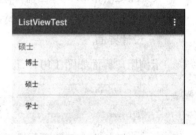

图 3.9 列表框

3.3 界面事件

在 Android 系统中,存在多种界面事件,如点击事件、触摸事件、焦点事件和菜单事件等,在这些事件发生时,Android 界面框架调用界面控件的事件处理函数对事件进行处理。

3.3.1 按键事件

在 MVC 模型中,控制器根据界面事件(UI Event)类型不同,将事件传递给界面控件不同的事件处理函数。例如,按键事件(KeyEvent)将传递给 onKey()函数进行处理,触摸事件(TouchEvent)将传递给 onTouch()函数进行处理。

Android 系统界面事件的传递和处理遵循一定的规则。

如果界面控件设置了事件监听器,则事件将先传递给事件监听器。如果界面控件没有设置事件监听器，界面事件则会直接传递给界面控件的其他事件处理函数。

即使界面控件设置了事件监听器，界面事件也可以再次传递给其他事件处理函数。如果监听器处理函数的返回值为 true，表示该事件已经完成处理过程，不需要其他处理函数参与处理过程，这样事件就不会再继续进行传递。否则事件会传递给其他的事件处理函数。

例如，假设 EditText 控件已经设置了按键事件监听器。当用户按下键盘上的某个按键时，控制器将产生 KeyEvent 按键事件。因为 EditText 控件已经设置按键事件监听器 OnKeyListener，所以按键事件先传递到监听器的事件处理函数 onKey()中。如果 onKey()函数返回 false，事件将继续传递，EditText 控件就可以捕获到该事件，将按键的内容显示在 EditText 控件中。如果 onKey()函数返回 true，将阻止按键事件的继续传递，这样 EditText 控件就不能够捕获按键事件，也就不能够将按键内容显示在 EditText 控件中。

Android 界面框架支持对按键事件的监听，并能够将按键事件的详细信息传递给处理函数。为了处理控件的按键事件，先需要设置按键事件的监听器，并重载 onKey()函数。示例代码如下。

```
entryText.setOnKeyListener(new OnKeyListener(){          //(a)
    @Override
    public boolean onKey(View view, int keyCode, KeyEvent keyEvent) {   //(b)
        //过程代码……
        return true ; //or false;                         //(c)
    }
})
```

其中：

（a）设置控件的按键事件监听器。

（b）第一个参数 view 表示产生按键事件的界面控件；第二个参数 keyCode 表示按键代码；第三个参数 keyEvent 则包含了事件的详细信息，如按键的重复次数、硬件编码和按键标志等。

（c）onKey()函数的返回值，返回 true，阻止事件传递；返回 false，允许继续传递按键事件。

下面通过一个例子说明 KeyEventTest 是一个说明如何处理按键事件。

1. 设计界面

示例用户界面如图 3.10 所示。

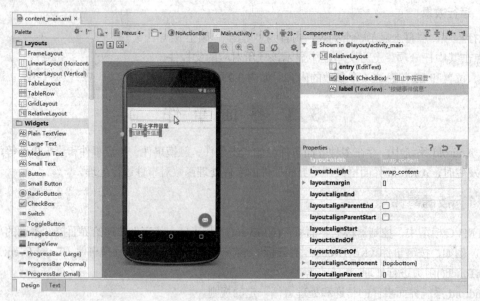

图 3.10 按钮界面

在 KeyEventTest 的用户界面中，最上方的 EditText 控件是输入字符的区域，中间的 CheckBox 控件用来控制 onKey()函数的返回值，最下方的 TextView 控件用来显示按键事件的详细信息，包括按键动作、按键代码、按键字符、Unicode 编码、重复次数、功能键状态、硬件编码和按键标志。

界面的 XML 文件的代码如下。

```xml
<?xml version="1.0" encoding="utf-8"?>
<RelativeLayout xmlns:android="http://schemas.android.com/apk/res/android"
    ...>

    <EditText
        android:layout_width="fill_parent"
        android:layout_height="wrap_content"
        android:id="@+id/entry"
        android:textSize="@dimen/abc_text_size_large_material"
        android:textColor="@android:color/black" />

    <CheckBox
        android:layout_width="wrap_content"
        android:layout_height="wrap_content"
        android:text="阻止字符回显"
        android:id="@+id/block"
        android:layout_below="@+id/entry"
        android:textSize="@dimen/abc_text_size_large_material" />

    <TextView
        android:layout_width="wrap_content"
        android:layout_height="wrap_content"
        android:text="按键事件信息"
        android:id="@+id/label"
        android:layout_below="@id/block"
        android:textSize="@dimen/abc_text_size_large_material" />
</RelativeLayout>
```

2. 处理事件

在 EditText 中，每当键盘任何一个键按下或抬起时都会引发按键事件。但为了能够使 EditText 处理按键事件，需要使用 setOnKeyListener()函数在代码中设置按键事件监听器，并在 onKey()函数添加按键事件的处理过程。

在 MainActivity.java 文件中，代码如下：

```java
package com.easybooks.android.keyeventtest;

import android.os.Bundle;
...

public class MainActivity extends AppCompatActivity {
    private EditText myentry;
    private CheckBox myblock;
    private TextView mylabel;

    @Override
    protected void onCreate(Bundle savedInstanceState) {
```

```
            super.onCreate(savedInstanceState);
            setContentView(R.layout.activity_main);
            findViews();
            Toolbar toolbar = (Toolbar) findViewById(R.id.toolbar);
    ...
            myentry.setOnKeyListener(new View.OnKeyListener() {
                @Override
                public boolean onKey(View v, int keyCode, KeyEvent event) {
                    int metaState = event.getMetaState();                              //(a)
                    int unicodeChar = event.getUnicodeChar();                          //(b)
                    String msg = "";
                    msg += "按键动作：" + String.valueOf(event.getAction()) + "\n";    //(c)
                    msg += "按键代码：" + String.valueOf(keyCode) + "\n";
                    msg += "按键字符：" + (char) unicodeChar + "\n";                   //(b)
                    msg += "UNICODE：" + String.valueOf(unicodeChar) + "\n";
                    msg += "重复次数：" + String.valueOf(event.getRepeatCount()) + "\n"; //(d)
                    msg += "功能键状态：" + String.valueOf(metaState) + "\n";          //(a)
                    msg += "硬件编码：" + String.valueOf(event.getScanCode()) + "\n";  //(e)
                    msg += "按键标志：" + String.valueOf(event.getFlags()) + "\n";     //(f)
                    mylabel.setText(msg);
                    if (myblock.isChecked()) return true;
                    return false;
                }
            });
        }

        private void findViews() {
            myentry = (EditText) findViewById(R.id.entry);
            myblock = (CheckBox) findViewById(R.id.block);
            mylabel = (TextView) findViewById(R.id.label);
        }
        ...
    }
```

其中：

（a）用来获取功能键状态。功能键包括左 Alt 键、右 Alt 键和 Shift 键，当这三个功能键被按下时，功能键代码 metaState 值分别为 18、34 和 65。没有功能键被按下时，功能键代码 metaState 值都为 0。

（b）获取按键的 Unicode 值。将 Unicode 转换为字符，显示在 TextView 中。

（c）获取按键动作，0 表示按下按键，1 表示抬起按键。

（d）获取按键的重复次数，但按键被长时间按下时，则会产生这个属性值。

（e）获取了按键的硬件编码，不同硬件设备的按键硬件编码都不相同，因此该值一般用于调试。

（f）获取了按键事件的标志符。

3. 运行效果

运行程序，按键盘上的 "A" 按键，下方显示出该按键事件的详细信息，同时上方的文本输入区也会显示出'a'字符；在勾选 "阻止字符回显" 复选框后，按键盘上的 "B" 按键，下方显示按键事件的信息，但上方却不会同时显示'b'字符，如图 3.11 所示。

实际进行 APP 开发时需要这些信息来判断，以进行进一步处理或者提示用户。

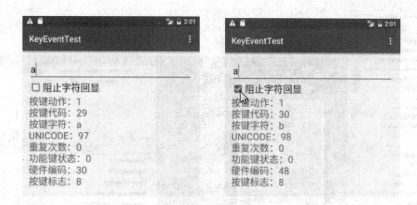

图 3.11 按键事件

3.3.2 触摸事件

伴随着触摸屏的普及,手机的操作方式也随之改变,用户已经不满足键盘的操作方式,而是更加倾心于使用手指在屏幕上进行操作。

Android 界面框架支持对触摸事件的监听,并能够将触摸事件的详细信息传递给处理函数。为了处理控件的触摸事件,首先需要设置触摸事件的监听器,并重载 onTouch()函数,示例代码如下:

```
touchView.setOnTouchListener(new View.OnTouchListener(){                //(a)
        @Override
        public boolean onTouch(View v, MotionEvent event) {              //(b)
            //过程代码……
            return true/false;                                           //(c)
        }
})
```

其中:

(a) 设置控件的触摸事件监听器。

(b) 第一个参数 View 表示产生触摸事件的界面控件,第二个参数 MotionEvent 是触摸事件的详细信息,如产生时间、坐标和触点压力等。

(c) onTouch()函数的返回值。

下面以一个示例说明如何处理触摸事件。

1. 设计界面

设计用户界面如图 3.12 所示。

在用户界面中,上方浅色区域是可以接收触摸事件的区域,用户在台式机上 APP 开发时可以在 Android 模拟器中使用鼠标点击屏幕,用以模拟触摸手机屏幕。下方黑色区域是显示区域,用来显示触摸事件的类型、相对坐标、绝对坐标、触点压力、触点尺寸和历史数据量等信息。

在用户界面包含三个 TextView 控件,第一个 TextView(ID 为 touch_area)用来标识触摸事件的测试区域,第二个 TextView(ID 为 history_label)用来显示触摸事件的历史数据量,第三个 TextView(ID 为 event_label)用来显示触摸事件的详细信息,包括类型、相对坐标、绝对坐标、触点压力和触点尺寸。

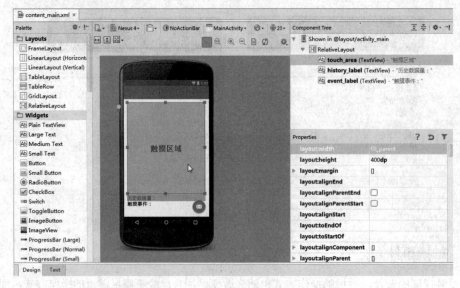

图 3.12 触摸手机界面

XML 文件的代码如下。

```xml
<?xml version="1.0" encoding="utf-8"?>
<RelativeLayout xmlns:android="http://schemas.android.com/apk/res/android"
    ...>

    <TextView
        android:layout_width="fill_parent"
        android:layout_height="400dp"
        android:text="触摸区域"
        android:id="@+id/touch_area"
        android:background="@color/material_deep_teal_200"           //(a)
        android:textSize="@dimen/abc_text_size_display_1_material"
        android:textStyle="bold"
        android:gravity="center" />

    <TextView
        android:layout_width="fill_parent"
        android:layout_height="wrap_content"
        android:text="历史数据量："
        android:id="@+id/history_label"
        android:textSize="@dimen/abc_text_size_large_material"
        android:layout_below="@id/touch_area"
        android:background="@android:color/darker_gray" />           //(a)

    <TextView
        android:layout_width="wrap_content"
        android:layout_height="wrap_content"
        android:id="@+id/event_label"
        android:textColor="@android:color/black"                     //(b)
        android:textSize="@dimen/abc_text_size_large_material"
        android:text="触摸事件："
        android:layout_below="@id/history_label"
```

```
            android:textStyle="bold" />
</RelativeLayout>
```

其中:

(a) 定义 TextView 的背景颜色。

(b) 定义 TextView 的字体颜色。

2. 处理事件

当手指接触到触摸屏、在触摸屏上移动或离开触摸屏时,分别会引发 ACTION_DOWN、ACTION_UP 和 ACTION_MOVE 触摸事件。而无论是哪种触摸事件,都会调用 onTouch()函数进行处理。事件类型包含在 onTouch()函数的 MotionEvent 参数中,可以通过 getAction()函数获取到触摸事件的类型,然后根据触摸事件的不同类型进行不同的处理。但为了能够使屏幕最上方的 TextView 处理触摸事件,需要使用 setOnTouchListener()函数在代码中设置触摸事件监听器,并在 onTouch()函数添加触摸事件的处理过程。

(1) 处理事件代码

在 MainActivity.java 文件中,代码如下:

```
package com.easybooks.android.toucheventtest;

import android.os.Bundle;
...

public class MainActivity extends AppCompatActivity {
    private TextView touchView;
    private TextView historyView;
    private TextView eventView;

    @Override
    protected void onCreate(Bundle savedInstanceState) {
        super.onCreate(savedInstanceState);
        setContentView(R.layout.activity_main);
        findViews();
        Toolbar toolbar = (Toolbar) findViewById(R.id.toolbar);
        ...
        touchView.setOnTouchListener(new View.OnTouchListener() {
            @Override
            public boolean onTouch(View v, MotionEvent event) {
                int action = event.getAction();
                switch (action) {
                    case MotionEvent.ACTION_DOWN:
                        display("触击屏幕", event);
                        break;
                    case MotionEvent.ACTION_UP:
                        processHistory(event);
                        display("离开屏幕", event);
                        break;
                    case MotionEvent.ACTION_MOVE:
                        display("在屏幕上移动", event);
                        break;
                }
                return true;
            }
```

```java
            });
        }

        private void findViews() {
            touchView = (TextView) findViewById(R.id.touch_area);
            historyView = (TextView) findViewById(R.id.history_label);
            eventView = (TextView) findViewById(R.id.event_label);
        }

        private void display(String eventType, MotionEvent event) {        //(a)
            int x = (int) event.getX();                                     //(b)
            int y = (int) event.getY();                                     //(b)
            int rawX = (int) event.getRawX();                               //(c)
            int rawY = (int) event.getRawY();                               //(c)
            float pressure = event.getPressure();                           //(d)
            float size = event.getSize();                                   //(e)

            String msg = "";
            msg += "事件类型: " + eventType + "\n";
            msg += "相对坐标: (" + String.valueOf(x) + ", " + String.valueOf(y) + ")\n";
            msg += "绝对坐标: (" + String.valueOf(rawX) + ", " + String.valueOf(rawY) + ")\n";
            msg += "触点压力: " + String.valueOf(pressure) + "\n";
            msg += "触点尺寸: " + String.valueOf(size) + "\n";
            eventView.setText(msg);
        }

        private void processHistory(MotionEvent event) {                    //(f)
            int historySize = event.getHistorySize();                       //(g)
            for (int i = 0; i < historySize; i++) {                         //(h)
                long time = event.getHistoricalEventTime(i);                //(i)
                float x = event.getHistoricalX(i);                          //(i)
                float y = event.getHistoricalY(i);                          //(j)
                float pressure = event.getHistoricalPressure(i);            //(k)
                float size = event.getHistoricalSize(i);                    //(l)
                //处理过程......
            }
            historyView.setText("历史数据量: " + historySize);              //(m)
        }
    ...
}
```

Display()函数说明：

（a）Display()是一个自定义函数，将 MotionEvent 参数中的事件信息提取出来，并显示在用户界面上。不仅有触摸事件的类型信息，还有触点的坐标信息。

（b）获取的是触点相对于父界面元素的坐标信息。

（c）获取的是触点绝对坐标的坐标信息。

（d）获取触点压力。触点压力是一个介于 0 和 1 之间的浮点数，表示用户对触摸屏施加压力的大小，接近 0 表示压力较小，接近 1 表示压力较大。

（e）获取触点尺寸。触点尺寸指用户接触触摸屏的接触点大小，也是一个介于 0 和 1 之间的浮点数，接近 0 表示尺寸较小，接近 1 表示尺寸较大。

（2）触摸屏事件

一般情况下，如果用户将手指放在触摸屏上，但不移动，然后抬起手指，应先后产生 ACTION_DOWN 和 ACTION_UP 两个触摸事件。但如果用户在屏幕上移动手指，然后再抬起手指，则会产生这样的事件序列 ACTION_DOWN→ACTION_MOVE→ACTION_MOVE→ACTION_MOVE→……→ACTION_UP。

在手机上运行的应用程序，效率是非常重要的。如果 Android 界面框架不能产生足够多的触摸事件，则应用程序就不能够很精确地描绘触摸屏上的触摸轨迹。相反，如果 Android 界面框架产生了过多的触摸事件，虽然能够满足精度的要求，但却降低了应用程序效率。Android 界面框架使用了"打包"的解决方法。在触点移动速度较快时会产生大量的数据，每经过一定的时间间隔便会产生一个 ACTION_MOVE 事件，在这个事件中，除了有当前触点的相关信息外，还包含这段时间间隔内触点轨迹的历史数据信息，这样既能够保持精度，又不至于产生过多的触摸事件。通常情况下，在 ACTION_MOVE 的事件处理函数中，都先处理历史数据，然后再处理当前数据。

ProcessHistory()函数说明：

（f）ProcessHistory()也是一个自定义函数，用来处理触摸事件的历史数据。
（g）获取历史数据的数量。
（h）循环处理这些历史数据。
（i）获取历史事件的发生时间。
（j）获取历史事件的触点压力。
（k）获取历史事件的相对坐标。
（l）获取历史事件的触点尺寸。
（m）界面显示历史数据的数量。

Android 模拟器并不支持触点压力和触点尺寸的模拟，所有触点压力恒为 1.0，触点尺寸恒为 0.0。同时，在 Android 模拟器上也无法产生历史数据，因此历史数据量一直显示为 0。

3. 运行效果

运行程序，用鼠标在模拟器屏幕上点击（模拟指尖触击），然后拖曳鼠标（模拟手指在屏幕上移动），最后释放鼠标（模拟手指离开触摸屏）的过程，界面下方显示出其间发生的事件类型及详细信息，如图 3.13 所示。

图 3.13 触摸事件

3.4 高级控件应用

3.4.1 网页浏览控件：WebView

WebView 能加载显示网页，可以将其视为一个浏览器。用于文字超链接和图片超链接非常简单，它仅仅帮助转向相应的网页，而显示网页是 Android 的相应功能完成，不再处于本 APP 的控制之下。而 WebView 主要功能本身也可作为浏览器客户端使用，它使用了 WebKit 渲染引擎加载显示网页。实现 WebView 的步骤如下。

（1）在 Activity 中实例化 WebView 组件：WebView webView = new WebView(this);
或者在布局文件中声明 WebView，在 Activity 中定义 WebView 控件变量。
（2）调用 WebView 的 loadUrl()方法，设置 WebView 要显示的网页：
webView.loadUrl("显示的网页 URL");
（3）调用 Activity 的 setWebView( )方法来显示网页视图。
（4）为了让 WebView 支持回退功能，覆盖系统 Activity 类的 onKeyDown()方法，否则浏览器会调用 finish()而结束自身，而不是回退到上一页面。
（5）需要在 AndroidManifest.xml 文件中添加权限：
<uses-permission android:name="android.permission.INTERNET" />
否则会出现 Web page not available 错误。

【例 3.7】设计一个网页进行图书查询。

1. 设计界面

创建工程 WebViewTest，选择"Basic Activity"。
设计界面如图 3.14 所示。

图 3.14　Web 视图

其中：用了3个WebView控件，第一行左边WebView控件（ID为WebImg）用于图片超链接，右边WebView控件（ID为WebText）用于文字超链接。下部一个大区域的WebView控件（ID为WebClient）作为浏览器在其中显示网页。

其他控件包括：

2个TextView控件：显示"图书查询功能"提示和"查询图书名"提示。

1个EditText控件（ID为editBook）：输入需要查询的图书名，默认为"计算机导论"。

1个Button控件（ID为buttonOK）："提交"查询命令。

在XML文件（content_main.xml）中，代码如下：

```xml
<?xml version="1.0" encoding="utf-8"?>
<RelativeLayout xmlns:android="http://schemas.android.com/apk/res/android"
    ...>

    <WebView                                    //(a)
        android:layout_width="265dp"
        android:layout_height="52dp"
        android:id="@+id/webImg"
        android:layout_alignParentTop="true"
        android:layout_alignParentLeft="true"
        android:layout_alignParentStart="true" />

    <WebView                                    //(a)
        android:layout_width="wrap_content"
        android:layout_height="wrap_content"
        android:id="@+id/webText"
        android:layout_alignBottom="@+id/webImg"
        android:layout_toRightOf="@+id/webImg"
        android:layout_toEndOf="@+id/webImg" />

    <TextView
        android:layout_width="wrap_content"
        android:layout_height="wrap_content"
        android:text="图书查询功能"
        android:id="@+id/textTitle"
        android:layout_below="@+id/webImg"
        android:layout_alignParentLeft="true"
        android:layout_alignParentStart="true"
        android:paddingTop="10dp"
        android:textColor="@android:color/black"
        android:textStyle="bold"
        android:textSize="@dimen/abc_text_size_large_material" />

    <TextView
        android:layout_width="wrap_content"
        android:layout_height="wrap_content"
        android:text="查询图书名："
        android:id="@+id/textView"
        android:layout_below="@+id/textTitle"
        android:layout_alignParentLeft="true"
        android:layout_alignParentStart="true"
```

```xml
        android:paddingTop="10dp"
        android:textColor="@android:color/black"
        android:textSize="@dimen/abc_text_size_medium_material" />

    <EditText
        android:layout_width="wrap_content"
        android:layout_height="wrap_content"
        android:id="@+id/editBook"
        android:layout_below="@+id/textTitle"
        android:layout_toRightOf="@+id/textView"
        android:layout_toEndOf="@+id/textView"
        android:text="MySQL 实用教程" />

    <Button
        android:layout_width="wrap_content"
        android:layout_height="wrap_content"
        android:text="提交"
        android:id="@+id/buttonOk"
        android:layout_alignBottom="@+id/editBook"
        android:layout_toRightOf="@+id/editBook"
        android:layout_toEndOf="@+id/editBook"
        android:textSize="@dimen/abc_text_size_medium_material"
        android:onClick="onOkClick" />                          //(b)

    <WebView                                                     //(a)
        android:layout_width="match_parent"
        android:layout_height="match_parent"
        android:id="@+id/webClient"
        android:layout_below="@+id/editBook"
        android:layout_alignParentRight="true"
        android:layout_alignParentEnd="true" />
</RelativeLayout>
```

其中:

(a) 定义了 3 个 WebView 控件,它们的差别仅仅在于控件的 ID 和位置属性。

(b) 定义了单击"提交"按钮对应的事件代码。

2. 功能实现

(1) 在当前工程 app\src\main\res\drawable 目录下存放图片资源:

db0.jpg, njnu.jpg

(2) 在 MainActivity.java 文件中,代码如下:

```java
package com.easybooks.android.webviewtest;

import android.os.Bundle;
...

public class MainActivity extends AppCompatActivity {
    private WebView myWebImg;                            //(a)
    private WebView myWebText;                           //(a)
    private WebView myWebClient;                         //(a)
    private final String mimeType = "text/html";
```

```java
        private final String encoding = "utf-8";

        @Override
        protected void onCreate(Bundle savedInstanceState) {
            super.onCreate(savedInstanceState);
            setContentView(R.layout.activity_main);
            findViews();
            loadWebPages();                                          //(b)
            Toolbar toolbar = (Toolbar) findViewById(R.id.toolbar);
            ...
            myWebClient.setWebViewClient(new WebViewClient() {       //(c)
                @Override
                public boolean shouldOverrideUrlLoading(WebView view, String url) {
                    view.loadUrl(url);
                    return true;
                }
            });
        }

        private void findViews() {
            myWebImg = (WebView) findViewById(R.id.webImg);
            myWebText = (WebView) findViewById(R.id.webText);
            myWebClient = (WebView) findViewById(R.id.webClient);
        }
        //(b)*****载入网页****
        private void loadWebPages() {
            myWebImg.loadDataWithBaseURL(null, "<a href='http://www.njnu.edu.cn'><img src='file:///android_res/drawable/njnu.jpg'/></a>", mimeType, encoding, null);          //(b.1)
            myWebText.loadDataWithBaseURL(null, "<a  href='http://computer.njnu.edu.cn'>计算机科学与技术学院</a>", mimeType, encoding, null);          //(b.2)
            myWebClient.getSettings().setJavaScriptEnabled(true);    //设置网页采用 Java 脚本
            myWebClient.loadUrl("http://www.hxedu.com.cn");          //(b.3)
        }
        //(c) *****载入网页****
        public void onOkClick(View view){                            //(d)
            myWebClient.loadDataWithBaseURL(null, "<a  href='http://www.hxedu.com.cn/hxedu/fg/book/bookinfo.html?code=G0232700'><img src='file:///android_res/drawable/db0.jpg'/></a>", mimeType, encoding, null);
        }

        @Override
        public boolean onKeyDown(int keyCode, KeyEvent event) {      //(e)
            if (keyCode == KeyEvent.KEYCODE_BACK &&myWebClient.canGoBack()) {
                myWebClient.goBack();
                return true;
            }
            return super.onKeyDown(keyCode, event);
        }
        ...
    }
```

其中：

（a）定义3个WebView控件变量。

（b）载入第一个 WebView 控件图片超链接；载入第二个 WebView 控件文字超链接；载入第三个 WebView 控件初始显示网页。

（c）对 webClient 控件重写 setWebViewClient 代码取代系统浏览网页方法，以在 webClient 控件显示网页。

（d）定义单击"提交"按钮的事件代码。单击"提交"按钮，显示本地目录下 res/drawable/db0.jpg 图片。

（e）为了让 webClient 控件支持回退功能，重写 onKeyDown()代码。

3. 配置 AndroidManifest.xml

在 AndroidManifest.xml 中添加配置（加黑处），以在文件中添加浏览网页权限。

代码如下：

```xml
<?xml version="1.0" encoding="utf-8"?>
<manifest xmlns:android="http://schemas.android.com/apk/res/android"
    package="com.easybooks.android.webviewtest">
    <uses-permission android:name="android.permission.INTERNET"/>

    <application
     ...
    </application>

</manifest>
```

4. 运行效果

运行程序，系统显示如图 3.15（a）所示。单击"南京师范大学"图片超链接，系统显示如图 3.15（b）所示。采用手机"后退"功能回退，再单击"计算机科学与技术学院"超链接，系统显示如图 3.15（c）所示。

图 3.15　初始页面点击图片和文字超链接

采用手机"后退"功能回退,再单击"提交"按钮,系统显示如图 3.16(a)所示。单击 MySQL 实用教程图片,系统显示如图 3.16(b)所示。

(a)

(b)

图 3.16 点击文字链接和"提交"按钮

3.4.2 滚动预览控件:HorizontalScrollView

HorizontalScrollView 是用于布局的容器,支持水平方向的滚动显示。

HorizontalScrollView 可通过其中放置一个 LinearLayout 子控件。然后向 LinearLayout 添加其他的控件,最后达到丰富其内容的效果。其中,LinearLayout 设置的 orientation 布局为 Horizontal。HorizontalScrollView 不可以和 ListView 同时用,因为 ListView 有自己的滚动条设置。

【例 3.8】设计一个应用,通过横向滚屏,显示图书封面图片,如图 3.17 所示。

图 3.17 滚动显示

1. 设计界面

创建工程 HorizontalScrollViewTest,选择"Basic Activity"。

加入 HorizontalScrollView 控件,在 HorizontalScrollView 控件中加入线性控件(LinearLayout)。在

运行时动态加入图书封面图片到 LinearLayout 中（LinearLayout 详细用法在后面一章介绍）。

设计界面如图 3.18 所示。

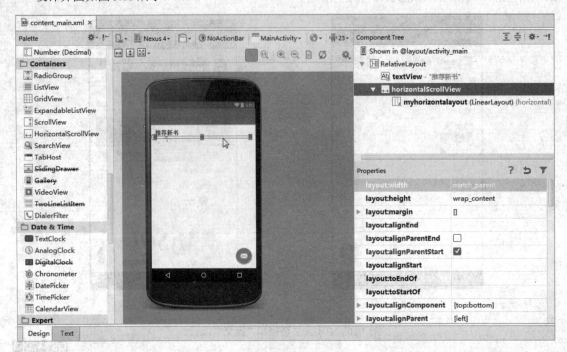

图 3.18 滚动界面设计

在 XML 文件（content_main.xml）中，代码如下：

```xml
<?xml version="1.0" encoding="utf-8"?>
<RelativeLayout xmlns:android="http://schemas.android.com/apk/res/android"
    ...>

    <TextView
        android:layout_width="wrap_content"
        android:layout_height="wrap_content"
        android:text="推荐新书"
        android:textSize="@dimen/abc_text_size_large_material"
        android:textColor="@android:color/black"
        android:id="@+id/textView" />

    <HorizontalScrollView
        android:layout_width="match_parent"
        android:layout_height="wrap_content"
        android:id="@+id/horizontalScrollView"
        android:layout_below="@+id/textView"
        android:layout_alignParentLeft="true"
        android:layout_alignParentStart="true"
        android:paddingTop="10dp">

        <LinearLayout
            android:orientation="horizontal"
            android:layout_width="match_parent"
```

```xml
            android:layout_height="match_parent"
            android:id="@+id/linearLayout"
            android:background="@android:color/holo_blue_dark">
</LinearLayout>
    </HorizontalScrollView>
</RelativeLayout>
```

在当前工程 app\src\main\res\drawable 目录下存放图片资源:

cp.jpg, db0.jpg, db2.jpg, db4.jpg, java0.jpg, java2.jpg, qt1.jpg, qt3.jpg, qt4.jpg

2. 功能实现

在 MainActivity.java 文件中,res\drawable 目录下存放图片资源的 ID 用数组保存,在 findViews() 实现将图片动态加入到 LinearLayout 中。

代码如下:

```java
package com.easybooks.android.horizontalscrollviewtest;
import android.os.Bundle;
...
public class MainActivity extends AppCompatActivity {
    private LinearLayout myLinearLayout;        //定义 LinearLayout 控件变量
    private int[] image = {
            R.drawable.cp, R.drawable.db0, R.drawable.db2, R.drawable.db4, R.drawable.java0, R.drawable.java2, R.drawable.qt1, R.drawable.qt3, R.drawable.qt4
    };                                          //定义图片变量数组,用图书封面图片资源赋值

    @Override
    protected void onCreate(Bundle savedInstanceState) {
        super.onCreate(savedInstanceState);
        setContentView(R.layout.activity_main);
        findViews();
        Toolbar toolbar = (Toolbar) findViewById(R.id.toolbar);
        ...
    }
    //(a)动态创建 linearLayout 中的图片显示
    private void findViews() {
        myLinearLayout = (LinearLayout) findViewById(R.id.linearLayout);   //对 linearLayout 控件引用
        for (Integer id : image) {                                          //遍历得到每一个图片 ID
            myLinearLayout.addView(addImage(id));                           //(b)
        }
    }
    //(b) 将 ID 对应图片加入 linearLayout 控件中
    private View addImage(Integer id) {
        LinearLayout layout1 = new LinearLayout(getApplicationContext());   //动态创建 LinearLayout 控件
        layout1.setLayoutParams(new LinearLayout.LayoutParams(293, 404));   //设置 LinearLayout 控件宽高
        layout1.setGravity(Gravity.CENTER);                                 //LinearLayout 中的图片对中
        ImageView imageView1 = new ImageView(getApplicationContext());      //动态创建 ImageView 控件
        imageView1.setLayoutParams(new LinearLayout.LayoutParams(273, 384));
        imageView1.setImageResource(id);            //将 ID 作为 ImageView 控件图片
        layout1.addView(imageView1);                //将 ImageView 控件加入到 LinearLayout 控件
        return layout1;
    }
    ...
}
```

3. 运行效果

运行程序，效果如图 3.17 所示。通过滚动可以显示其他图片。

3.4.3 照片查看器：ImageSwitcher

ImageSwitcher 控件用于实现类似于 Windows 操作系统下的"Windows 照片查看器"中的上一张、下一张切换图片的功能。

ImageSwitcher 有两个子 View：ImageView，当左右滑动的时候，就在这两个 ImageView 之间来回切换来显示图片。在使用 ImageSwitcher 时，必须实现 ViewSwitcher.ViewFactory 接口，并通过 makeView()方法来创建用于显示图片的 ImageView。makeView()方法将返回一个显示图片的 ImageView。在使用图片切换器时，setImageResource()方法用于指定要在 ImageSwitcher 中显示的图片资源。

【例 3.9】设计一个应用，通过左右按钮，显示图书封面图片，如图 3.19 所示。

1. 设计界面

创建工程 ImageSwitcherTest，选择"Basic Activity"。

（1）在当前工程 app\src\main\res\drawable 目录下除了下列图片文件：

图 3.19 图像查看

cp.jpg、db0.jpg、db2.jpg、db4.jpg、java0.jpg、java2.jpg、qt1.jpg、qt3.jpg、qt4.jpg 另外增加：backward.png 和 forward.png 文件。

（2）界面上放 ImageSwitcher 控件，另外上面放一个 TextView，在其左右放 2 个图片按钮显示图片分别为 backward.png 和 forward.png。

设计界面如图 3.20 所示。

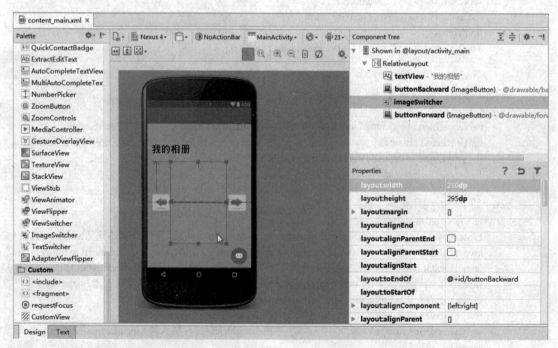

图 3.20 图像查看界面设计

在 XML 文件（content_main.xml）中，代码如下：

```xml
<?xml version="1.0" encoding="utf-8"?>
<RelativeLayout xmlns:android="http://schemas.android.com/apk/res/android"
    ...
    android:background="@android:color/darker_gray">

<TextView
    android:layout_width="wrap_content"
    android:layout_height="wrap_content"
    android:text="我的图书"
    android:textColor="@android:color/black"
    android:textSize="@dimen/abc_text_size_display_1_material"
    android:id="@+id/textView"
    android:paddingBottom="25dp"
    android:textStyle="bold"
    android:layout_above="@+id/imageSwitcher"
    android:layout_centerHorizontal="true" />

<ImageButton
    android:layout_width="wrap_content"
    android:layout_height="wrap_content"
    android:id="@+id/buttonBackward"
    android:src="@drawable/backward"
    android:onClick="onClick"                        //单击按钮事件调用 onClick 方法
    android:layout_alignBottom="@+id/imageSwitcher"
    android:layout_alignParentLeft="true"
    android:layout_alignParentStart="true" />

<ImageSwitcher
    android:layout_width="210dp"
    android:layout_height="295dp"
    android:id="@+id/imageSwitcher"
    android:layout_centerVertical="true"
    android:layout_toRightOf="@+id/buttonBackward"
    android:layout_toEndOf="@+id/buttonBackward" />

<ImageButton
    android:layout_width="wrap_content"
    android:layout_height="wrap_content"
    android:id="@+id/buttonForward"
    android:layout_alignTop="@+id/buttonBackward"
    android:layout_toRightOf="@+id/imageSwitcher"
    android:layout_toEndOf="@+id/imageSwitcher"
    android:src="@drawable/forward"
    android:onClick="onClick" />                     //单击按钮事件调用 onClick 方法

</RelativeLayout>
```

2. 功能实现

在 MainActivity.java 文件中，代码如下：

```
package com.easybooks.android.imageswitchertest;
import android.os.Bundle;
```

```java
    ...
    public class MainActivity extends AppCompatActivity implements ViewSwitcher.ViewFactory {
        private ImageSwitcher myImageSwitcher;                //定义 ImageSwitcher 控件变量
        private int[] image = {
            R.drawable.cp, R.drawable.db0, R.drawable.db2, R.drawable.db4, R.drawable.java0, R.drawable.java2,
            R.drawable.qt1, R.drawable.qt3, R.drawable.qt4 };          //图书数组存放图片资源 ID
        int index;                                            //当前图片的索引（从 0 开始）

        @Override
        protected void onCreate(Bundle savedInstanceState) {
            super.onCreate(savedInstanceState);
            setContentView(R.layout.activity_main);
            findViews();
            Toolbar toolbar = (Toolbar) findViewById(R.id.toolbar);
            ...
        }

        private void findViews() {
            myImageSwitcher = (ImageSwitcher) findViewById(R.id.imageSwitcher);    //赋值控件变量引用
            myImageSwitcher.setFactory(this);                                      //与工厂联系
            myImageSwitcher.setImageDrawable(getResources().getDrawable(image[0])); //显示第 1 幅图
            index = 0;
        }

        @Override
        public View makeView() {                                                   //显示图片
            ImageView imageView1 = new ImageView(this);                            //创建 ImageView 控件
            imageView1.setBackgroundColor(0x00000000);                             //设置背景颜色
            imageView1.setScaleType(ImageView.ScaleType.FIT_CENTER);
            imageView1.setLayoutParams(new ImageSwitcher.LayoutParams(
                ViewGroup.LayoutParams.MATCH_PARENT, ViewGroup.LayoutParams.MATCH_PARENT));
            return imageView1;
        }
        //图片按钮单击事件
        public void onClick(View view) {
            switch (view.getId()) {
                case R.id.buttonBackward:                                          //向后箭头按钮按键处理
                    index--;
                    if (index < 0) {
                        index = image.length - 1;
                        Toast.makeText(this, "温馨提示：您正在查看最后一张图片", Toast.LENGTH_SHORT).show();  //显示系统提示信息对话框
                    }
                    //将 Index 顺序对应图片 ID 与显示图片联系
                    myImageSwitcher.setImageResource(image[index]);
                    break;
                case R.id.buttonForward:                                           //向前箭头按钮按键处理
                    index++;
                    if (index >= image.length) {
                        index = 0;
                        Toast.makeText(this, "温馨提示：您正在查看第一张图片", Toast.LENGTH_SHORT).show();
```

```
            myImageSwitcher.setImageResource(image[index]);
            break;
        }
    }
    ...
}
```

说明：

（1）显示一组图片采用工厂机制。所以需要首先实现接口：

```
public class MainActivity extends AppCompatActivity implements ViewSwitcher.ViewFactory {
    ...
}
```

然后可以使用：

myImageSwitcher.setFactory(this);   //与工厂联系

（2）重写 ImageSwitcher 控件 makeView()方法，以按照本题要求显示图片。

其中：

采用 imageView 控件图片显示。

"imageView1.setScaleType(ImageView.ScaleType.FIT_CENTER);"语句按图片的原来 size 居中显示，当图片长/宽超过 View 的长/宽，则截取图片的居中部分显示。

"imageView1.setLayoutParams(new ImageSwitcher.LayoutParams(
ViewGroup.LayoutParams.MATCH_PARENT, ViewGroup.LayoutParams.MATCH_PARENT))"：利用 setLayoutParams 在代码中调整布局。ViewGroup 相当于一个放置 View 的容器，给子 View 计算出建议的宽和高与测量模式（这里是"填充满"父容器）。

（3）图片显示

findViews() { … }中显示第 1 张图片；单击左右按钮，更换显示图片。具体通过单击事件中的下列语句更换 ImageSwitcher 控件的图片资源实现：

myImageSwitcher.setImageResource(image[index]);

3. 运行效果

运行程序，首先显示 cp.jpg 图片，单击向右箭头按钮 4 次，显示 java0.jpg，结果如图 3.19 所示。因为 img[0]和 img[4]分别是 cp.jpg 和 java0.jpg 图片资源。

3.4.4 条类控制器：SeekBar/RatingBar

（1）SeekBar

SeekBar 称为拖动条，可以被用户控制。对其进行事件监听，需要实现 setOnSeekBarChangeListener 接口。在 SeekBar 中需要监听 3 个事件，分别是：

① 数值的改变：onProgressChanged，我们可以得到当前数值的大小。
② 开始拖动：onStartTrackingTouch。
③ 停止拖动：onStopTrackingTouch。

（2）RatingBar

RatingBar 称为等级条，用星形来显示等级评定。对其进行事件监听，需要实现 setOnRatingBarChangeListener()接口。

星值的改变事件：onRatingChanged()。

在使用默认 RatingBar 时,用户可以通过触摸/拖动/按键(比如遥控器)来设置评分。

【例 3.10】设计一个应用,通过拖动条,缩放图书封面图片。通过等级条评定等级,如图 3.21 所示。

1. 设计界面

创建工程 BarTest,选择"Basic Activity"。

设计界面如图 3.22 所示。

其中:

(1) seekBar 控件

设置 seekBar 控件,实现图片放大和缩小。本例我们考虑的对应关系如下。

图 3.21 条类控件

显示图片:	38%	50%	100%
进程:	0	12(初始值)	62(max)

说明:

① 设置 SeekBar 控件属性:android:max="62"。
② setProgress(12)方法可以将 SeekBar 值设置为 12。
③ scaleMin=38,图片缩放比例:(scaleMin+拖动条值)/100。

拖动条为 0;38%。

初始状态:(12+38=50)%。

任意位置:(scaleMin+ getProgress())/100,其中 getProgress()方法可以得到当前的位置值。

(2) RatingBar 控件

① 设置 RatingBar 控件属性:android:numStars="5",设置 5 颗星。
② SetRatingBar(3) 方法设置 RatingBar 控件为 3 颗星。

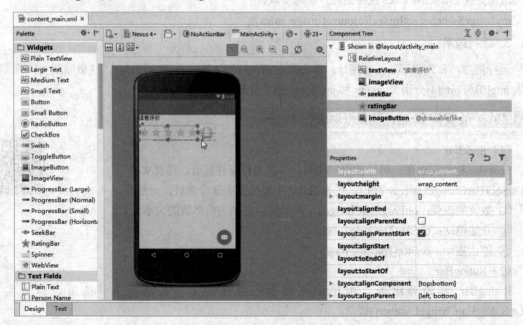

图 3.22 条类控件界面

③ 通过 getRatingBar()方法可得到星的值。

在 XML 文件（content_main.xml）中，代码如下：

```xml
<?xml version="1.0" encoding="utf-8"?>
<RelativeLayout xmlns:android="http://schemas.android.com/apk/res/android"
    ...
    android:background="@color/dim_foreground_disabled_material_dark">

    <TextView
        android:layout_width="wrap_content"
        android:layout_height="wrap_content"
        android:text="读者评价"
        android:textColor="@android:color/black"
        android:textSize="@dimen/abc_text_size_large_material"
        android:id="@+id/textView" />

    <ImageView
        android:layout_width="wrap_content"
        android:layout_height="wrap_content"
        android:id="@+id/imageView"
        android:layout_below="@+id/textView"
        android:layout_alignParentLeft="true"
        android:layout_alignParentStart="true"
        android:layout_toLeftOf="@+id/imageButton"
        android:layout_toStartOf="@+id/imageButton" />

    <SeekBar
        android:layout_width="225dp"
        android:layout_height="wrap_content"
        android:id="@+id/seekBar"
        android:layout_below="@+id/imageView"
        android:layout_alignParentLeft="true"
        android:layout_alignParentStart="true"
        android:max="62" />

    <RatingBar
        android:layout_width="wrap_content"
        android:layout_height="wrap_content"
        android:id="@+id/ratingBar"
        android:layout_alignParentLeft="true"
        android:layout_alignParentStart="true"
        android:layout_below="@+id/seekBar"
        android:layout_alignParentBottom="false"
        android:numStars="5" />

    <ImageButton
        android:layout_width="wrap_content"
        android:layout_height="wrap_content"
        android:id="@+id/imageButton"
        android:src="@drawable/like"
        android:layout_alignBottom="@+id/ratingBar"
```

```xml
            android:layout_toRightOf="@+id/ratingBar"
            android:layout_toEndOf="@+id/ratingBar"
            android:onClick="onButtonClick" />

</RelativeLayout>
```

在项目 app\src\main\res\drawable 目录下存放资源：db2.jpg 和 like.jpg。

2. 功能实现

在 MainActivity.java 文件中，代码如下：

```java
package com.easybooks.android.bartest;

import android.graphics.Bitmap;
...

public class MainActivity extends AppCompatActivity {
    private ImageView myImageView;
    private SeekBar mySeekBar;
    private RatingBar myRatingBar;
    private TextView myTextView;
    private Bitmap myBitmap;
    private int width, height;
    private final int scaleMin = 38;                  //作为 SeekBar 滑条的最小值

    @Override
    protected void onCreate(Bundle savedInstanceState) {
        super.onCreate(savedInstanceState);
        setContentView(R.layout.activity_main);
        findViews();
        Toolbar toolbar = (Toolbar) findViewById(R.id.toolbar);
...
        //监听界面 SeekBar 控件
        mySeekBar.setOnSeekBarChangeListener(new SeekBar.OnSeekBarChangeListener() {
            @Override
            public void onProgressChanged(SeekBar seekBar, int progress, boolean fromUser) {
                scaleToFit((float) (seekBar.getProgress() + scaleMin) / 100);     //监听 SeekBar 数值改变
            }

            @Override
            public void onStartTrackingTouch(SeekBar seekBar) {                   //监听 SeekBar 开始

            }

            @Override
            public void onStopTrackingTouch(SeekBar seekBar) {                    //监听 SeekBar 停止

            }
        });
        //监听界面 RatingBar 控件
        myRatingBar.setOnRatingBarChangeListener(new RatingBar.OnRatingBarChangeListener() {
            @Override
            public void onRatingChanged(RatingBar ratingBar, float rating, boolean fromUser) {
```

```java
            myTextView.setText("读者评价: " + rating + "星");                //监听RatingBar星值改变
        }
    });
}

private void findViews() {
    myImageView = (ImageView) findViewById(R.id.imageView);
    mySeekBar = (SeekBar) findViewById(R.id.seekBar);
    mySeekBar.setProgress(12);                      //设置拖动条初始位置
    myRatingBar = (RatingBar) findViewById(R.id.ratingBar);
    myRatingBar.setRating(3);                       //初始显示3颗星
    myTextView = (TextView) findViewById(R.id.textView);
    myBitmap = BitmapFactory.decodeResource(getResources(), R.drawable.db2);
    width = myBitmap.getWidth();                    //获得原图片的宽度
    height = myBitmap.getHeight();                  //获得原图片的高度
    scaleToFit((float) (12 + scaleMin) / 100);      //显示图片初始大小为原图的50%/(12+38)%
}
//使图片按照（scale*100）%显示
public void scaleToFit(float scale) {
    Matrix matrix = new Matrix();                   //创建临时矩阵
    matrix.postScale(scale*2, scale*2);             //设置临时矩阵大小，*2为放大显示
    Bitmap bmp = Bitmap.createBitmap(myBitmap, 0, 0, width, height, matrix, true);  //创建缩放后图片
    myImageView.setImageBitmap(bmp);                //在图片控件中显示
}
//按钮单击事件
public void onButtonClick(View view) {
    myRatingBar.setRating(myRatingBar.getRating() + 0.5f);       //单击一次按钮，加0.5颗星
}
    ...
}
```

3. 运行效果

运行程序，拖动条到最大，单击3次按钮，显示如图3.21所示。

第 4 章 用户界面布局

用户通过界面使用 APP，在 Android 中通过界面布局组件对界面进行设计。

4.1 界面布局

界面布局（Layout）是对用户界面结构的描述，定义了界面中控件的相互关系。

1. 界面布局的含义

Android 界面布局一般在 XML 文件中描述，这样可以更好地分离程序的表现层和控制层，在仅仅修改界面时不再需要更改程序的源代码。通过可视化工具直接设计的用户界面，为界面设计带来了极大的便利。

但在程序运行时，仍然可以通过代码动态添加或修改界面布局，更新状态等。

2. 界面布局的类型

Android 界面布局包括 LinearLayout（线性布局）、RelativeLayout（相对布局）、TableLayout（表格布局）、GridLayout（网格布局）、AbsoluteLayout（绝对布局）和 FrameLayout（版块布局）。

在创建 APP 工程的过程中，在"Add an Activity to Mobile"对话框用户选择"Basic Activity"，系统创建的 content_main.xml 界面布局文件默认采用相对布局（RelativeLayout），用户创建的新的界面布局文件可以选择布局的类型。

3. 创建界面布局文件

创建新的界面布局文件有以下两种方法。

（1）选择"layout"，按右键，选择 File→New→Layout resource file，如图 4.1 所示。

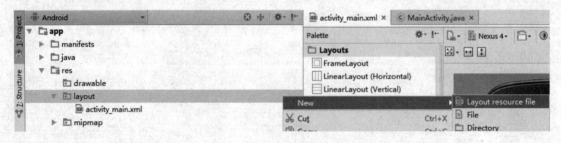

图 4.1 创建新的 layout 文件

此后系统显示创建界面文件的对话框，例如输入有关的内容：

File name（界面文件名）：mylayout
Root element（根元素）：LinearLayout

输入其他内容后单击"OK"，系统创建了 mylayout.xml 界面布局文件，自动保存到 res\layout 文件夹下，如图 4.2 所示。

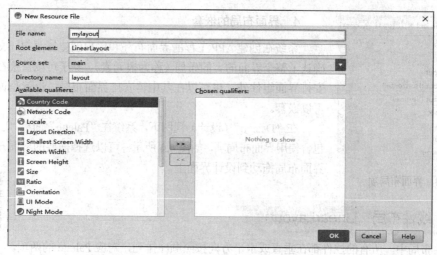

图 4.2　Layout resource file 设置

这种方法在创建界面布局时还可设置很多属性。

（2）在 Android 开发环境中，选择主菜单 File→New→XML→Layout XML File 项。系统显示创建界面文件的对话框，例如输入有关的内容：

Layout File name：mylayout
Root Tag：LinearLayout

输入其他内容后单击"OK"，系统创建了 mylayout.xml 界面布局文件，如图 4.3 所示。

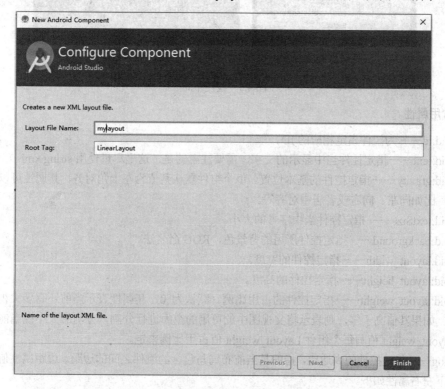

图 4.3　创建界面布局

单击"Finish"按钮，创建 mylayout.xml 文件，自动保存到 res\layout 文件夹下。

4. 界面布局的嵌套

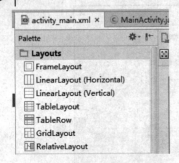

图 4.4 界面布局项

系统在创建 APP 工程的界面布局和用户创建界面布局文件时仅仅确认了整个界面的大的布局（称为根元素）。除了可以直接在该布局下安排界面组件外，也可以在其中再安排界面布局。也就是说界面布局可以嵌套。

在"Design"（设计）视图下，系统在"Palette"面板"Layouts"下包含使用界面布局项，如图 4.4 所示。可以选择需要的界面布局或者把界面布局拖动到设计界面上。

4.1.1 线性布局：LinearLayout

在线性布局中，所有的控件都在垂直或水平方向按照顺序排列，对应 Palette 的两项。

（1）LinearLayout(horizontal)项：android:orientation="horizontal"，包含控件水平排列，每行仅包含一个界面控件。如图 4.5（a）所示。

（2）LinearLayout(vertical)项：android:orientation="vertical"，包含控件垂直排列，每列仅包含一个界面控件。如图 4.5（b）所示。

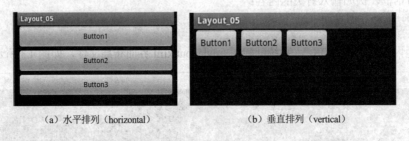

(a) 水平排列（horizontal）   (b) 垂直排列（vertical）

图 4.5 线性布局

1．常用属性

android:id——为控件指定相应的 ID。

android:text——指定控件当中显示的文字，需要注意的是，这里尽量使用 string.xml。

android:gravity——指定控件的基本位置。每个组件默认其值为左上角对齐，其属性可以调整组件对齐方式，比如向左、向右或者居中对齐等。

android:textSize——指定控件当中字号的大小。

android:background——指定控件所用的背景色，RGB 命名法。

android:layout_width——指定控件的宽度。

android:layout_height——指定控件的高度。

android:layout_weight——指定控件的占用比例。默认为 0，其属性表示当前还有多大视图就占据多大视图。如果其值高于零，则表示将父视图中的可用的空间进行分割，分割的大小视当前屏幕整体布局的 Layout_weight 值与每个组件 Layout_weight 值占用比例来定。

padding——指定控件的内边距，指的是当前布局与包含的组件之间的边距。边距属性值为具体数字。其上下左右属性如下。

android:paddingTop——设置上边距。

android:paddingBottom——设置下边距。

android:paddingLeft——设置左边距。
android:paddingRight——设置右边距。
android:padding——设置周围四方向各内边距统一调整。
layout_margin——指定控件的外边距,是与其他组件之间的边距。其上下左右属性如下。
android:layout_marginTop——设置上边距。
android:layout_marginBottom——设置下边距。
android:layout_marginLeft——设置左边距。
android:layout_marginRight——设置右边距。
android:layout_margin——设置四方向外边距统一调整。
android:singleLine——如果设置为真的话,则将控件的内容显示在一行当中。

2. 线性布局嵌套

一般来说,线性布局需要嵌套才能多行布局。例如,用线性布局完成图 4.6 的界面。

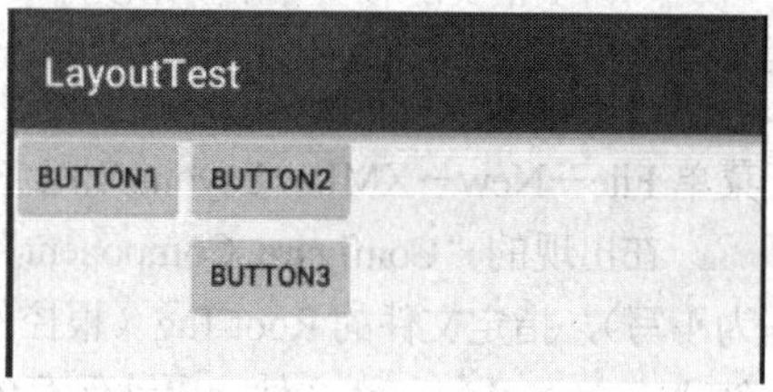

图 4.6 一个线性布局里头嵌套另一个线性布局

界面布局代码如下:

```xml
<?xml version="1.0" encoding="utf-8"?>
<LinearLayout xmlns:android="http://schemas.android.com/apk/res/android"
    android:layout_width="match_parent"
    android:layout_height="match_parent"
    android:orientation="horizontal" >

    <Button
        android:layout_width="wrap_content"
        android:layout_height="wrap_content"
        android:text="Button1"
        />
    <LinearLayout
        android:orientation="vertical"
        android:layout_width="fill_parent"
        android:layout_height="fill_parent"
        >
        <Button
            android:layout_width="wrap_content"
            android:layout_height="wrap_content"
            android:text="Button2"
            />
        <Button
            android:layout_width="wrap_content"
            android:layout_height="wrap_content"
            android:text="Button3"
            />

    </LinearLayout>
</LinearLayout>
```

【例 4.1】两种线性布局如图 4.7 所示,试设计实现该布局的界面。

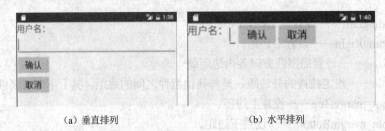

(a) 垂直排列　　　　　　　　　　(b) 水平排列

图 4.7　水平排列和垂直排列的线性布局

（1）建立一个新的 Android 工程，工程名称为 LinearLayoutTest。在 Android 开发环境中，选择主菜单 File→New→XML→Layout XML File 项，打开 Layout（布局）类型的 XML 文件建立向导。

在出现的"Configure Component"对话框中，为 Layout 文件命名为"mainvertical"（注意必须全为小写），指定文件的 Root Tag（根控件标签）为"LinearLayout"（线性布局），单击"Finish"按钮，创建 mainvertical.xml 文件，自动保存到 res\layout 下。

双击新建立的 res\layout\mainvertical.xml 文件，打开线性布局的属性编辑器。线性布局的排列方法由 orientation 属性控制，vertical 表示垂直排列，horizontal 表示水平排列。这里 orientation 属性的值选择 vertical。修改线性布局的属性：

```
android:layout_width="wrap_content"
android:layout_height="wrap_content"
```

（2）用户按照 TextView、EditText、Button、Button 的顺序，将 4 个界面控件先后拖曳到可视化编辑器中，所有控件会自动按拖曳的顺序显示在可视化编辑器中。将界面控件位置确定后，按照表 4.1 在属性编辑器中修改界面控件的属性。

表 4.1　线性布局界面控件的属性设置

类　型	属　性	值
TextView	id	@+id/label
	Text	用户名：
EditText	id	@+id/entry
	layout_width	fill_parent
	text	""
Button	id	@+id/ok
	text	确认
Button	id	@+id/cancel
	text	取消

（3）打开 XML 文件编辑器，查看 mainvertical.xml 文件代码。

代码如下：

```xml
<?xml version="1.0" encoding="utf-8"?>
<LinearLayout xmlns:android="http://schemas.android.com/apk/res/android"
    android:layout_width="fill_parent"
    android:layout_height="wrap_content"
    android:baselineAligned="false"
    android:orientation="vertical">

    <TextView
        android:layout_width="wrap_content"
```

```
        android:layout_height="wrap_content"
        android:text="用户名："
        android:id="@+id/label"
        android:textSize="@dimen/abc_text_size_large_material" />

    <EditText
        android:layout_width="fill_parent"
        android:layout_height="wrap_content"
        android:text=""
        android:id="@+id/entry"
        android:textSize="@dimen/abc_text_size_large_material" />

    <Button
        android:layout_width="wrap_content"
        android:layout_height="wrap_content"
        android:text="确认"
        android:id="@+id/ok"
        android:textSize="@dimen/abc_text_size_large_material" />

    <Button
        android:layout_width="wrap_content"
        android:layout_height="wrap_content"
        android:text="取消"
        android:id="@+id/cancel"
        android:textSize="@dimen/abc_text_size_large_material" />
</LinearLayout>
```

（4）将 activity_main.xml 文件中的 "<include layout="@layout/content_main" />" 更改为 "<include layout="@layout/mainvertical" />"，如图 4.8 所示。

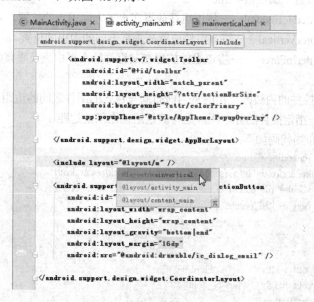

图 4.8　修改包含的布局文件

运行后的结果见图 4.7（a）。

建立横向排列的线性布局过程与上述的纵向垂直线性布局相似，效果见图 4.7（b）。

4.1.2 相对布局：RelativeLayout

相对布局是一种非常灵活的布局方式，是 Android Studio 默认的布局方式。它通过指定当前控件与对应 ID 值控件的相对位置关系，确定界面中所有控件的布局位置。和线性布局一样，它能够最大程度地保证在各种屏幕类型的手机上正确显示界面布局。

常用属性如下。

(1) 组件之间的位置关系

android:layout_above——将组件放在指定 ID 组件的上方。

android:layout_below——将组件放在指定 ID 组件的下方。

android:layout_toLeftOf——将组件放在指定 ID 组件的左方。

android:layout_toRightOf——将组件放在指定 ID 组件的右方。

(2) 组件对齐方式

android:layout_alignBaseline——将该组件放在指定 ID 组件进行中心线对齐。

android:layout_alignTop——将该组件放在指定 ID 组件进行顶部对齐。

android:layout_alignBottom——将该组件放在指定 ID 组件进行底部对齐。

android:layout_alignLeft——将该组件放在指定 ID 组件进行左边缘对齐。

android:layout_alignRight——将该组件放在指定 ID 组件进行右边缘对齐。

(3) 当前组件与父组件的对齐方式，属性值="true"

android:layout_alignParentTop——该组件与父组件进行顶部对齐。

android:layout_alignParentBottom——该组件与父组件进行底部对齐。

android:layout_alignParentLeft——该组件与父组件进行左边缘对齐。

android:layout_alignParentRight——该组件与父组件进行右边缘对齐。

(4) 组件放置的位置，属性值="true"

android:layout_centerHorizontal——将该组件放置在水平方向中央的位置。

android:layout_centerVertical——将该组件放置在垂直方向的中央的位置。

android:layout_centerInParent——将该组件放置在父组件的水平中央及垂直中央的位置。

(5) 其他

padding——指定控件的内边距，指的是当前布局与包含的组件之间的边距。

layout_margin——指定控件的外边距，是与其他组件之间的边距。

例如，相对界面布局代码如下：

```
<?xml version="1.0" encoding="utf-8"?>
<RelativeLayout xmlns:android="http://schemas.android.com/apk/res/android"
    android:layout_width="fill_parent"
    android:layout_height="fill_parent"
    >

    <Button
        android:layout_width="wrap_content"
        android:layout_height="wrap_content"
        android:text="Button1"
        android:id="@+id/btn1"
        />
    <Button
```

```
            android:layout_width="fill_parent"
            android:layout_height="wrap_content"
            android:text="Button2"
            android:id="@+id/btn2"
            android:layout_below="@id/btn1"
            />
    <Button
            android:layout_width="wrap_content"
            android:layout_height="wrap_content"
            android:text="Button3"
            android:id="@+id/btn3"
            android:layout_below="@id/btn2"
            android:layout_alignRight="@id/btn2"
            />
    <Button
            android:layout_width="wrap_content"
            android:layout_height="wrap_content"
            android:text="Button4"
            android:id="@+id/btn4"
            android:layout_below="@id/btn3"
            android:layout_alignParentRight="true"
            />
    <Button
            android:layout_width="wrap_content"
            android:layout_height="wrap_content"
            android:text="Button5"
            android:id="@+id/btn5"
            android:layout_below="@id/btn4"
            android:layout_centerHorizontal="true"
            />
</RelativeLayout>
```

其中：

Button1 在界面左上角，Button2 在 Button1 下面，Button3 在 Button2 下面并且对齐 Button2 右边缘，Button4 在 Button3 下面并且对齐父组件右边缘，Button5 在 Button4 下面并且在水平方向的中央位置。

相对界面布局的结果如图 4.9 所示。

实际操作时为了方便起见，用上列相对界面布局代码取代前面的线性布局代码进行试验。

【例 4.2】用相对布局设计如图 4.10 所示界面。

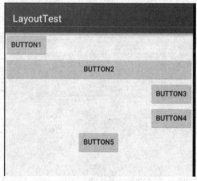

图 4.9　相对布局

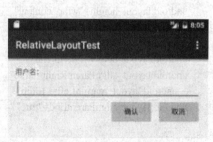

图 4.10　相对布局界面

Android 实用教程（基于 Android Studio·含视频分析）

（1）建立一个新的 Android 工程，工程名称为 RelativeLayoutTest，在界面可视化编辑器上拖曳添加控件，工程默认采用的就是相对布局。

（2）添加 TextView 控件（"用户名"），相对布局会将 TextView 控件放置在屏幕的最上方。添加 EditText 控件（输入框），并声明该控件的位置在 TextView 控件的下方，相对布局会根据 TextView 的位置确定 EditText 控件的位置。添加第一个 Button 控件作为"取消"按钮，声明在 EditText 控件的下方，且在父控件的最右边。添加第二个 Button 控件作为"确认"按钮，声明该控件在第一个 Button 控件的左方，且与第一个 Button 控件处于相同的水平位置。

（3）打开 XML 文件编辑器，查看 content_main.xml 文件代码。

代码如下：

```xml
<?xml version="1.0" encoding="utf-8"?>
<RelativeLayout xmlns:android="http://schemas.android.com/apk/res/android"        //(a)R-Up
    xmlns:app="http://schemas.android.com/apk/res-auto"
    xmlns:tools="http://schemas.android.com/tools"
    android:layout_width="match_parent"
    android:layout_height="match_parent"
    android:paddingBottom="@dimen/activity_vertical_margin"
    android:paddingLeft="@dimen/activity_horizontal_margin"
    android:paddingRight="@dimen/activity_horizontal_margin"
    android:paddingTop="@dimen/activity_vertical_margin"
    app:layout_behavior="@string/appbar_scrolling_view_behavior"
    tools:context="com.easybooks.android.relativelayouttest.MainActivity"
    tools:showIn="@layout/activity_main">                                         //(a)R-Down

    <TextView
        android:layout_width="wrap_content"
        android:layout_height="wrap_content"
        android:text="用户名："
        android:id="@+id/label"
        android:layout_alignParentTop="true"
        android:layout_alignParentLeft="true"
        android:layout_alignParentStart="true" />
    <EditText android:id="@+id/entry"
        android:layout_height="wrap_content"
        android:layout_width="fill_parent"
        android:layout_below="@+id/label">                                        //(b)
    </EditText>

    <Button
        android:layout_width="wrap_content"
        android:layout_height="wrap_content"
        android:text="取消"
        android:id="@+id/cancel"
        android:layout_below="@+id/entry"
        android:layout_alignParentRight="true"                                    //(c)
        android:layout_marginLeft="10dip"                                         //(d)
        android:layout_alignParentEnd="true" />

    <Button
        android:layout_width="wrap_content"
```

```
            android:layout_height="wrap_content"
            android:text="确认"
            android:id="@+id/ok"
            android:layout_below="@+id/entry"
            android:layout_toLeftOf="@+id/cancel"                    //(e)
            android:layout_alignTop="@id/cancel"                     //(f)
            android:layout_toStartOf="@+id/cancel" />

</RelativeLayout>
```

其中：

（a）使用了<RelativeLayout>标签声明一个相对布局。

（b）使用位置属性 android:layout_below，确定 EditText 控件在 ID 为 label 的控件下方。

（c）使用属性 android:layout_alignParentRight，声明该控件与其父控件的右边边界对齐。

（d）使用属性 android:layout_marginLeft，将该控件向左移动 10dip。

（e）声明使用属性 android:layout_toLeftOf，声明该控件在 ID 为 cancel 控件的左边。

（f）使用属性 android:layout_alignTop，声明该控件与 ID 为 cancel 的控件在相同的水平位置。

运行程序，效果如图 4.10 所示。

4.1.3 表格布局：TableLayout

表格布局也是一种常用的界面布局，它将屏幕划分为表格，通过指定行和列可以将界面控件添加到表格中。

表格布局 TableLayout 用<TableRow >……</TableRow>来表示一行，有多少行就添加多少个行标签对。有多少列就看最多的一个行中添加了多少个控件，直到把屏幕占满，超出屏幕的就不再显示。

直接在 TableLayout 加控件，控件会占据一行。

常用属性如下。

shrinkColumns 属性：当 TableRow 里面的控件布满布局时，指定列自动延伸以填充可用部分。当 TableRow 里面的控件还没有布满布局时，shrinkColumns 不起作用。

strechColumns 属性：指定列对空白部分进行填充。

collapseColumns 属性：隐藏指定的列。

layout_column 属性：设置组件显示指定列。

layout_span 属性：设置组件显示占用的列数。

注意，列号都是从 0 开始，收缩、拉伸或隐藏可以指定多个列，列之间必须用逗号隔开。例如，android:stretchColumns = "1,2"。也可以用"*"代替所有列。

各种表格布局界面如图 4.11 所示。

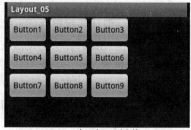 

　　　（a）一个 3 行 3 列表格　　　　　　　（b）指定第三列填充可用部分

图 4.11　各种表格布局

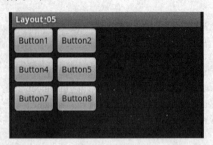

（c）指定第三列填充空白部分　　　　　（d）指定第三列隐藏

图4.11　各种表格布局（续）

其中：

（b）android:shrinkColumns="2"。

（c）android:strechColumns="2"。

（d）android:collapseColumns="2"。

例如，界面表格布局代码如下：

```xml
<?xml version="1.0" encoding="utf-8"?>
<TableLayout xmlns:android="http://schemas.android.com/apk/res/android"
    android:layout_width="fill_parent"
    android:layout_height="fill_parent">
    <TableRow>
        <Button android:text="Button1"
            android:layout_span="3"/>
        <Button android:text="Button2"/>
        <Button android:text="Button3"/>
    </TableRow>
    <TableRow>
        <Button android:text="Button4"
            android:layout_column="2"/>
        <Button android:text="Button5"
            android:layout_column="0"/>
        <Button android:text="Button6"/>
    </TableRow>
    <TableRow>
        <Button android:text="Button7"/>
        <Button android:text="Button8"/>
        <Button android:text="Button9"/>
    </TableRow>
</TableLayout>
```

其中：

Button1 被设置成占用了 3 列，Button4 被设置显示在第三列，但代码指定 Button5 显示在第一列，但没有按照设定显示，这样可知 TableRow 在表格布局中，一行里的组件都会自动放在前一组件的右侧，依次排列，只要确定了所在列，其后面的组件就无法再设置位置。

表格界面布局的结果如图 4.12 所示。

表格布局还支持嵌套，可以将一个表格布局放置在另一个表格布局的网格中，也可以在表格布局中添加其他界面布局。

【例4.3】使用表格布局实现用户界面，如图 4.13 所示。

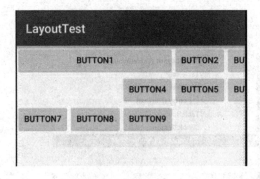
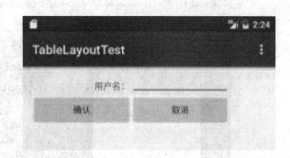

图 4.12　表格布局　　　　　　　　　　　　图 4.13　表格布局

（1）建立一个新的 Android 工程，工程名称为 TableLayoutTest。往设计界面上拖曳一个 TableLayout 控件，id 属性为 myTableLayout。TableLayout 是专用于表格布局的控件，当向其中拖入其他控件时，它自动呈现为网格状，如图 4.14 所示，即被分为很多行，每行又被划分为很多小的单元，每个单元中都可以添加一个界面控件。

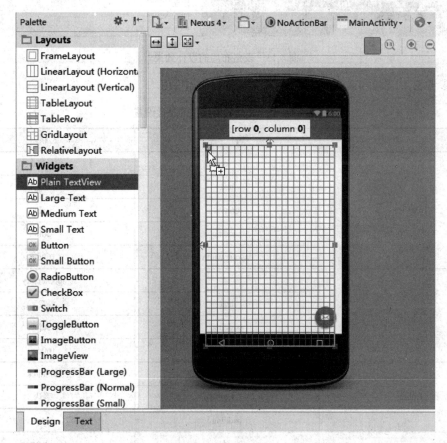

图 4.14　表格布局界面

在界面可视化编辑器上，向 TableLayout 的第一行前两个小格子中拖曳 TextView 和 EditText，再向第二行前两个小格子中分别拖曳两个 Button。实现效果如图 4.15 所示。

Android 实用教程（基于 Android Studio·含视频分析）

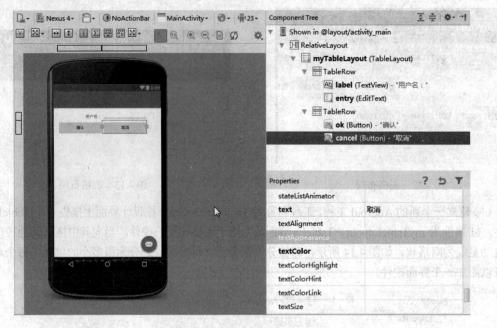

图 4.15 表格布局的设计效果

（2）参考表 4.2 设置 TableLayout 中 4 个界面控件的属性值。

表 4.2 表格布局界面控件的属性设置

类 型	属 性	值
TextView	id	@+id/label
	text	用户名：
	gravity	right
	padding	3dip
	layout_width	160dip
EditText	id	@+id/entry
	text	""
	padding	3dip
	layout_width	160dip
Button	id	@+id/ok
	text	确认
	padding	3dip
Button	id	@+id/cancel
	text	取消
	padding	3dip

（3）打开 XML 文件编辑器，查看 content_main.xml 文件代码。
代码如下：

```xml
<?xml version="1.0" encoding="utf-8"?>
<RelativeLayout xmlns:android="http://schemas.android.com/apk/res/android"
    xmlns:app="http://schemas.android.com/apk/res-auto"
    xmlns:tools="http://schemas.android.com/tools"
    android:layout_width="match_parent"
    android:layout_height="match_parent"
    android:paddingBottom="@dimen/activity_vertical_margin"
    android:paddingLeft="@dimen/activity_horizontal_margin"
    android:paddingRight="@dimen/activity_horizontal_margin"
    android:paddingTop="@dimen/activity_vertical_margin"
    app:layout_behavior="@string/appbar_scrolling_view_behavior"
    tools:context="com.easybooks.android.tablelayouttest.MainActivity"
    tools:showIn="@layout/activity_main">

    <TableLayout                                                              //(a)
        android:layout_width="match_parent"
        android:layout_height="match_parent"
        android:layout_alignParentTop="true"
        android:layout_alignParentLeft="true"
        android:layout_alignParentStart="true"
        android:id="@+id/myTableLayout">

        <TableRow                                                             //(b)
            android:layout_width="match_parent"
            android:layout_height="match_parent">

            <TextView
                android:text="用户名："
                android:id="@+id/label"
                android:layout_column="0"
                android:gravity="right"                                        //(c)
                android:padding="3dip"                                         //(d)
                android:layout_width="160dip" />                               //(e)

            <EditText
                android:layout_height="wrap_content"
                android:textAppearance="?android:attr/textAppearanceSmall"
                android:id="@+id/entry"
                android:layout_column="1"
                android:padding="3dip"
                android:layout_width="160dip" />
        </TableRow>

        <TableRow                                                             //(b)
            android:layout_width="match_parent"
            android:layout_height="match_parent">

            <Button
                android:text="确认"
                android:id="@+id/ok"
                android:layout_column="0"
```

```xml
                    android:padding="3dip"/>

                <Button
                    style="?android:attr/buttonStyleSmall"
                    android:layout_width="wrap_content"
                    android:layout_height="wrap_content"
                    android:text="取消"
                    android:id="@+id/cancel"
                    android:layout_column="1"
                    android:padding="3dip"/>
        </TableRow>
    </TableLayout>
</RelativeLayout>
```

其中：

（1）<TableLayout…></TableLayout>标签声明表格布局。

（2）<TableRow…></TableRow>代码声明了 TableRow 控件。共 2 个，用来表示布局中的两行。

（3）android:gravity="right"：将 TextView 中的文字对齐方式指定为右对齐。

（4）android:padding="3dip"：声明 TextView 控件与其他控件的间隔距离为 3dip。

（5）android:layout_width="160dip"：将 TextView 控件的宽度指定为 160dip。

运行程序，效果如图 4.13 所示。

4.1.4 网格布局：GridLayout

网格布局是 Android SDK 4.0（API Level 14）开始支持的布局方式，它将用户界面划分为网格，界面控件可随意摆放在这些网格中。一个界面控件可以占用多个网格（可以跨多行也可以跨多列），而在表格布局中却只能指定在一个表格行中，而不能跨越多个行。

网格布局的边界对用户是不可见的。在界面设计器中可以看到虚线网格，但在模拟器的运行结果中是看不到的。网格布局的块可以根据界面控件动态划分。

常用属性：

android:rowCount：设置网格布局行数。

android:columnCount：设置网格布局列数。

android:layout_row：设置组件位于第几行。

android:layout_column：设置组件位于第几列。

android:layout_rowSpan：设置组件纵向横跨几行。

android:layout_columnSpan：设置组件横向横跨几列。

网格布局和其他布局不同，可以不为组件设置 Layout_width 和 Layout_height 属性，因为组件的宽高由几行几列决定了，当然，也可以写 wrap_content。

【例 4.4】用程序实例演示网格布局的界面。

1. 创建网格布局

建立一个新的 Android 工程，工程名称为 GridLayoutTest，在界面可视化编辑器上拖曳放置一个 GridLayout 控件，再往其中拖曳添加控件，设计界面如图 4.16 所示。

打开 XML 文件编辑器，查看 content_main.xml 文件代码。

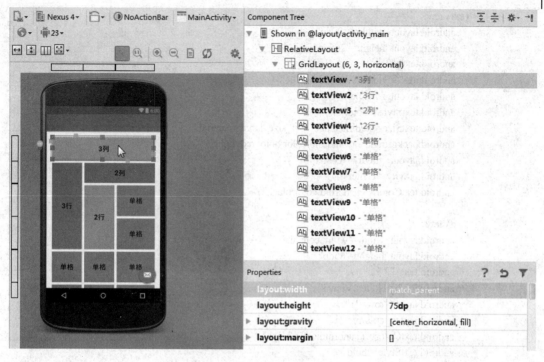

图4.16 网格布局界面

界面布局代码如下：

```xml
<?xml version="1.0" encoding="utf-8"?>
<RelativeLayout xmlns:android="http://schemas.android.com/apk/res/android"    //(a)R-Up
    ……
    tools:showIn="@layout/activity_main">                                       //(a)R-Down

    <GridLayout
        android:layout_width="match_parent"
        android:layout_height="match_parent"
        android:layout_alignParentTop="true"
        android:layout_centerHorizontal="true"
        android:useDefaultMargins="true"                                        //(b)
        android:rowCount="6"                                                    //(b)
        android:columnCount="3">                                                //(b)

        <TextView
            android:layout_height="75dp"
            android:text="3 列"
            android:id="@+id/textView"
            android:layout_columnSpan="3"                                       //(c)
            android:textSize="@dimen/abc_text_size_large_material"
            android:textStyle="bold"
            android:layout_gravity="center_horizontal|fill"
            android:background="@android:color/holo_blue_bright"
            android:gravity="center"
            android:textColor="@android:color/black" />
```

```xml
<TextView
    android:layout_width="100dp"
    android:layout_height="265dp"
    android:text="3 行"
    android:id="@+id/textView2"
    android:layout_rowSpan="3"
    android:textStyle="bold"
    android:textSize="@dimen/abc_text_size_large_material"
    android:background="@android:color/holo_red_light"
    android:layout_gravity="center|fill"                                //(d)
    android:gravity="center"
    android:textColor="@android:color/black" />

<TextView
    android:layout_width="wrap_content"
    android:layout_height="65dp"
    android:text="2 列"
    android:id="@+id/textView3"
    android:layout_row="1"                                              //(e)
    android:layout_column="1"                                           //(e)
    android:textColor="@android:color/black"
    android:textStyle="bold"
    android:textSize="@dimen/abc_text_size_large_material"
    android:gravity="center"
    android:layout_gravity="center|fill"
    android:background="@android:color/holo_green_dark"
    android:layout_columnSpan="2" />                                    //(e)

<TextView
    android:layout_width="100dp"
    android:layout_height="200dp"
    android:text="2 行"
    android:id="@+id/textView4"
    android:textColor="@android:color/black"
    android:textStyle="bold"
    android:textSize="@dimen/abc_text_size_large_material"
    android:gravity="center"
    android:layout_gravity="center|fill"
    android:background="@android:color/holo_orange_dark"
    android:layout_rowSpan="2" />                                       //(f)

<TextView
    android:layout_width="wrap_content"
    android:layout_height="100dp"
    android:text="单格"
    android:id="@+id/textView5"
    android:textColor="@android:color/black"
    android:textSize="@dimen/abc_text_size_large_material"
    android:gravity="center"
    android:layout_gravity="center|fill"
    android:background="@android:color/holo_purple"
```

```
                android:textStyle="bold" />

            <TextView
                android:layout_width="wrap_content"
                android:layout_height="100dp"
                android:text="单格"
                android:id="@+id/textView6"
                android:textColor="@android:color/black"
                android:textStyle="bold"
                android:textSize="@dimen/abc_text_size_large_material"
                android:gravity="center"
                android:layout_gravity="center|fill"
                android:background="@android:color/holo_purple"
                android:layout_row="3"                                      //(g)
                android:layout_column="2" />                                //(g)

            <TextView                                                       //(h)
                android:layout_width="wrap_content"
                android:layout_height="100dp"
                android:text="单格"
                android:id="@+id/textView7"
                android:layout_gravity="center|fill"
                android:textColor="@android:color/black"
                android:textStyle="bold"
                android:textSize="@dimen/abc_text_size_large_material"
                android:gravity="center"
                android:background="@android:color/holo_purple" />

            <TextView                                                       //(h)
                android:id="@+id/textView8"
                ……                                     />
            <TextView                                                       //(h)
                android:id="@+id/textView9"
                ……                                     />
            <TextView                                                       //(h)
                android:id="@+id/textView10"
                ……                                     />
            <TextView                                                       //(h)
                android:id="@+id/textView11"
                ……                                     />
            <TextView                                                       //(h)
                android:id="@+id/textView12"
                ……                                     />
            <TextView                                                       //(h)
                android:id="@+id/textView13"
                ……                                     />
    </GridLayout>
</RelativeLayout>
```

其中：

（a）使用了<RelativeLayout>标签声明一个相对布局。

（b）网格布局中的所有控件都遵循默认的边缘规则，就是说所有控件之间都会留有一定的边界空间，网格包含 6 行 3 列。

（c）TextView 控件占据 3 列。

（d）文字内容在所占据的块中居中显示。

（e）第二行第二列开始的 2 列。

（f）起始位置根据网格布局界面控件的排布规则，如果没有明确说明控件所在的块，那么当前控件会放置在前一个控件的同一行右侧的块上。如果前一个控件已经是这一行的末尾块，则当前控件放置在下一行的第一个块上。如果当前控件在纵向上占据多个块，而前一个控件右侧没有足够数量的块，则当前控件的起始位置也会放置在下一行的第一个块上。

（g）表示当前控件列的起始位置。如果所指定的列的位置在当前行已经被占用，则当前控件也会放置在下一行的这一列中。

（h）不指定位置，就是上一个控件的后一个位置。

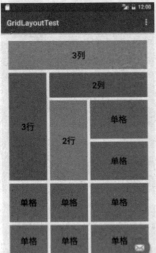

图 4.17　网格布局界面

2．运行效果

运行程序，效果如图 4.17 所示。

4.1.5　绝对布局：AbsoluteLayout

绝对布局能通过指定每一个界面控件的坐标位置。每一个界面控件都必须指定坐标（X，Y），坐标原点（0，0）在屏幕的左上角。

绝对布局是一种不推荐使用的界面布局，因为通过绝对位置确定的界面控件，虽然在特定目标手机上非常完美，但 Android 系统不能够根据不同屏幕大小对界面控件的位置进行调整，导致在其他不同类型的手机上，界面布局会变得非常混乱。

例如，一个绝对布局界面如图 4.18 所示。

界面布局代码如下：

```xml
<?xml version="1.0" encoding="utf-8"?>
<AbsoluteLayout
    xmlns:android="http://schemas.android.com/apk/res/android"
        android:orientation="vertical"
        android:layout_width="fill_parent"
        android:layout_height="fill_parent"
        >
    <Button
        android:layout_width="wrap_content"
        android:layout_height="fill_parent"
        android:text="Button1"
        android:layout_x="100dp"
        />
    <Button
        android:layout_width="fill_parent"
        android:layout_height="wrap_content"
        android:text="Button2"
        android:layout_y="100dp"
        />
```

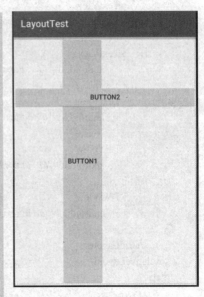

图 4.18　绝对布局界面

</AbsoluteLayout>

4.1.6 版块布局：FrameLayout

版块布局是最简单的界面布局，因为版块布局在新定义组件的时候都会将组件放置在屏幕的左上角，即使在此布局中定义多个组件，后一个组件总会将前一个组件所覆盖，除非最后一个组件是透明的。

属性如下。

android:foreground——设置帧布局容器的前景图像。前景图像永远处于帧布局最上面，直接面对用户的图像，就是不会被覆盖的图片。

android:foregroundGravity——设置前景图像显示的位置。

例如，一个版块布局界面如图 4.19 所示。

界面布局代码如下：

```xml
<?xml version="1.0" encoding="utf-8"?>
<FrameLayout xmlns:android="http://schemas.android.com/apk/res/android"
    android:layout_width="fill_parent"
    android:layout_height="fill_parent"
    >
    <Button
        android:layout_width="fill_parent"
        android:layout_height="wrap_content"
        android:text="Button1"
        />
    <Button
        android:layout_width="wrap_content"
        android:layout_height="fill_parent"
        android:text="Button2"
        />
    <Button
        android:layout_width="wrap_content"
        android:layout_height="wrap_content"
        android:text="Button3"
        />
</FrameLayout>
```

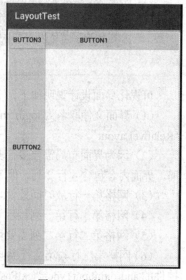

图 4.19　版块布局界面

在这个布局中，Button1 放在第一行，Button2 放在第一列，遮挡了 Button1 与 Button2 的相同部分。Button3 放在第一行第一列的 1 块，又遮挡了 Button2 与其相同的部分。

4.2　用户界面综合实例

4.2.1　【例一】：登录界面

创建 Android 工程，工程名称为 LoginPage，仍然选择 "Basic Activity"。

1. 设计页面

打开设计登录界面如图 4.20 所示。

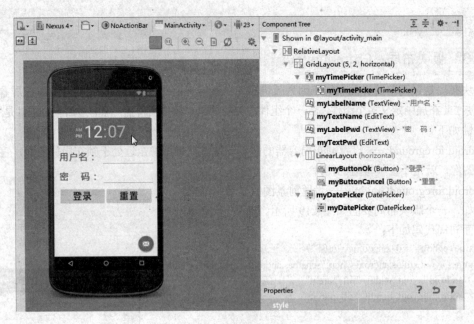

图 4.20　设计视图

可视化界面设计说明如下。

（1）界面文件取名为 login.xml，界面布局采用如图 4.1 所示方法创建，在"Root Element"中输入"RelativeLayout"。

（2）因为界面布局需要多行多列，而每一行包含的列不相同，大小不一致，所以创建一个网格布局。界面内容包含 4 行 2 列，但为了界面下面留空行，所以设置 5 行 2 列。

（3）网格第一行放时间控件显示时间，占满一行，所以需要合并 2 列。

（4）网格第二行第一列显示"用户名："，放 TextView 控件。

（5）网格第二行第二列需要用户输入用户名，放 EditText 控件。

（6）网格第三行与第二行类似，但用户输入密码时不能显示密码字符本身，所以需要设置 EditText 控件的 android:inputType="textPassword"。

（7）网格第四行需要放置 2 个命令按钮，希望分别控制占用宽度。所以需要在本行加入水平线性布局。

在线性布局<LinearLayout…>中设置：android:weightSum="1"。

在"登录"和"重置"命令按钮<Button…>中设置：android:layout_weight="0.5"。

通过调整 layout_weight 属性值可以改变命令按钮的宽度。

（8）网格第四行后面放置日期控件，在后续章节使用本实例时需要用到获取日期，因为不需要显示出来，所以需要设置 android:visibility="invisible"。

用户设计界面系统生成的 login.xml 文件，代码如下：

```xml
<?xml version="1.0" encoding="utf-8"?>
<RelativeLayout xmlns:android="http://schemas.android.com/apk/res/android"
    android:layout_width="match_parent" android:layout_height="match_parent">

    <GridLayout                                                    //(2)
        android:layout_width="match_parent"
        android:layout_height="match_parent"
```

```xml
    android:layout_alignParentTop="true"
    android:layout_centerHorizontal="true"
    android:useDefaultMargins="true"
    android:rowCount="5"
    android:columnCount="2">

    <TimePicker                                         //(3)
        android:layout_width="wrap_content"
        android:layout_height="99dp"
        android:id="@+id/myTimePicker"
        android:layout_columnSpan="2"/>

    <TextView                                           //(4)
        android:layout_width="wrap_content"
        android:layout_height="wrap_content"
        android:text="用户名："
        android:id="@+id/myLabelName"
        android:textStyle="bold"
        android:textSize="@dimen/abc_text_size_display_1_material"
        android:layout_row="1"
        android:layout_column="0" />

    <EditText                                           //(5)
        android:layout_width="wrap_content"
        android:layout_height="wrap_content"
        android:inputType="textPersonName"
        android:ems="10"
        android:id="@+id/myTextName"
        android:textSize="@dimen/abc_text_size_display_1_material"
        android:layout_row="1"
        android:layout_column="1" />

    <TextView                                           //(6)
        android:layout_width="wrap_content"
        android:layout_height="wrap_content"
        android:text="密    码："
        android:id="@+id/myLabelPwd"                    //(6)
        android:textStyle="bold"
        android:textSize="@dimen/abc_text_size_display_1_material"
        android:layout_row="2"
        android:layout_column="0" />

    <EditText                                           //(6)
        android:layout_width="wrap_content"
        android:layout_height="wrap_content"
        android:inputType="textPassword"
        android:ems="10"
        android:id="@+id/myTextPwd"
        android:textSize="@dimen/abc_text_size_display_1_material"
        android:layout_row="2"
        android:layout_column="1" />
```

```xml
            <LinearLayout                                    //(7)
                android:orientation="horizontal"
                android:layout_width="match_parent"
                android:layout_height="wrap_content"
                android:layout_columnSpan="2"
                android:weightSum="1"                        //(7)
                android:gravity="center">

                <Button
                    android:layout_width="wrap_content"
                    android:layout_height="wrap_content"
                    android:text="登录"
                    android:id="@+id/myButtonOk"
                    android:textSize="@dimen/abc_text_size_display_1_material"
                    android:textStyle="bold"
                    android:layout_weight="0.5"/>            //(7)

                <Button
                    android:layout_width="wrap_content"
                    android:layout_height="wrap_content"
                    android:text="重置"
                    android:id="@+id/myButtonCancel"
                    android:textSize="@dimen/abc_text_size_display_1_material"
                    android:textStyle="bold"
                    android:layout_weight="0.5"/>            //(7)
            </LinearLayout>

            <DatePicker                                      //(8)
                android:layout_width="wrap_content"
                android:layout_height="wrap_content"
                android:id="@+id/myDatePicker"
                android:visibility="invisible"               //(8)
                android:layout_columnSpan="2"/>

        </GridLayout>
</RelativeLayout>
```

2. 启动配置

一般情况下，运行时系统在 activity_main.xml 文件中查找下列项目：

```
<include layout="@layout/content_main" />
```

因为在创建当前工程时选择"Basic Activity"，它指定默认的运行界面为 content_main.xml 文件。因为当前首先运行的用户界面文件并没有放在 content_main.xml 文件，而是存放在 login.xml 中。所以需要修改 activity_main.xml 文件中的下列内容：

```
<include layout="@layout/login" />
```

3. 运行效果

运行效果如图 4.21 所示。

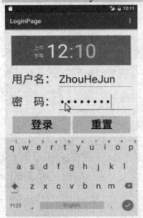

图 4.21　登录界面运行效果

4.2.2 【例二】：注册界面

设计注册 APP，效果如图 4.22 所示。

图 4.22 注册界面运行效果及日期对话框选择

其中：

用户名：直接输入。

密码和重输确认：直接输入，但显示"."。

性别：单击显示内容在"男"和"女"之间切换。

出生日期：单击后，显示日期对话框进行选择。

学历：在"博士"、"硕士"和"学士"列表中选择。

在"已阅读并接受网站服务条款"上打钩，"提交"按钮才能可用。

1. 创建 Android 工程

创建 Android 工程，工程名称为"RegisterPage"，选择"Empty Activity"。系统仅仅生成了"activity_main.xml"文件和 MainActivity.java 文件。

2. 设计页面

选择"File"→"New"→"XML"→"Layout XML File"，Layout File Name 输入"register"，修改 Root Tag：TableLayout。单击"Finish"，系统生成表格布局作为最外面布局的 register.xml 文件。

按照 APP 界面要求，直接编写 register.xml 文件代码如下：

```
<?xml version="1.0" encoding="utf-8"?>
<TableLayout xmlns:android="http://schemas.android.com/apk/res/android"
    android:layout_width="match_parent"
    android:layout_height="match_parent"
    android:layout_alignParentTop="true"
    android:layout_centerHorizontal="true">

    <TableRow>
        <TextView  android:textSize="@dimen/abc_text_size_headline_material"
```

```xml
        </TextView>
    </TableRow>
    <TableRow
        android:layout_gravity="center"
        android:gravity="center">
        <TextView
            android:text="请填写注册表单"
            android:textColor="@android:color/holo_blue_dark"
            android:textSize="@dimen/abc_text_size_headline_material"
            android:textStyle="bold" />
    </TableRow>
    <TableRow>
        <TextView
            android:text="用　户　名："
            android:textSize="@dimen/abc_text_size_large_material"
            android:gravity="right" />
        <EditText
            android:inputType="textPersonName"
            android:id="@+id/myEditName"
            android:gravity="left"  />
    </TableRow>
    <TableRow>
        <TextView
            android:text="密　　　码："
            android:textSize="@dimen/abc_text_size_large_material"
            android:gravity="right" />

        <EditText
            android:inputType="numberPassword"                        //(a)
            android:id="@+id/myEditPwd"
            android:hint="必须全部为数字"                               //(a)
            android:gravity="left" />
    </TableRow>
    <TableRow>
        <TextView
            android:text="重输确认："
            android:textSize="@dimen/abc_text_size_large_material"
            android:gravity="right" />
        <EditText
            android:inputType="numberPassword"                        //(a)
            android:id="@+id/myEditRePwd"
            android:gravity="left"  />
    </TableRow>
    <TableRow>
        <TextView
            android:text="性　　　别："
            android:textSize="@dimen/abc_text_size_large_material"
            android:gravity="right" />
        <ToggleButton
            android:id="@+id/myToggleButtonSex"
            android:textOn="男"                                       //(b)
```

```xml
            android:textOff="女"                                    //(b)
            android:checked="true" />                               //(b)
    </TableRow>
    <TableRow>
        <TextView
            android:text="出生日期："
            android:textSize="@dimen/abc_text_size_large_material"
            android:gravity="right" />
        <EditText
            android:inputType="date"                                //(c)
            android:id="@+id/myEditBirth"
            android:gravity="left" />
    </TableRow>
    <TableRow>
        <TextView
            android:text="学        历："
            android:textSize="@dimen/abc_text_size_large_material"
            android:gravity="right" />
        <Spinner
            android:id="@+id/mySpinnerDegree"/>                     //(d)
    </TableRow>
    <TableRow>
        <CheckBox
            android:text="已阅读并接受\n 网站服务条款"                  //(e)
            android:id="@+id/myCheckBoxAccept"
            android:layout_gravity="center|right"                   //(e)
            android:textStyle="bold"
            android:onClick="onCheckBoxClick" />                    //(e)
        <Button
            style="?android:attr/buttonStyleSmall"                  //(f)
            android:text="提        交"
            android:id="@+id/myButtonSubmit"
            android:textSize="@dimen/abc_text_size_large_material"
            android:enabled="false"                                 //(f)
            android:gravity="left"
            android:textAlignment="center" />
    </TableRow>
</TableLayout>
```

其中：

（a）文本编辑框输入内容均显示"."，初始提示信息为"必须全部为数字"。

（b）ToggleButton 控件属性 checked="true"（默认），显示"男"；属性 checked="false"，显示"女"。

（c）文本显示内容为日期。

（d）Spinner 控件选择输入内容，选择项目"博士、硕士、学士"在 findViews()中添加。

（e）提示信息包含"\n"表示换行；上下对中水平对右；单击按钮，执行 onCheckBoxClick 方法。

（f）命令按钮采用 buttonStyleSmall 风格，初始状态不可用。在 myCheckBoxAccept 控件的单击事件中判断两次密码相同，使该命令按钮可用。

3. 功能实现

创建 RegisterActivity.java 文件。在当前工程→app→java 下选择

com.easybooks.android.registerpage

按右键，选择 New→Java Class，系统显示如图 4.23 所示对话框：在 Name 处输入"RegisterActivity"。

RegisterActivity.java 设计思路如下。

（1）通过下列代码：

```
myBirth.setOnClickListener(new View.OnClickListener() {
    @Override
    public void onClick(View view) {
        …
    })
}
```

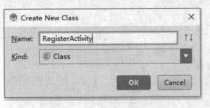

图 4.23　创建 java 类

监听日期输入控件单击事件，在其中通过 DatePickerDialog 控件显示日期选择对话框，以带入初始作为初始日期。通过 DatePickerDialog.OnDateSetListener DateListener 监听日期选择对话框的单击按钮事件，把日期对应的年月日放到相应的变量中，同时连接成日期字符串放入日期控件中。

（2）控件初始化方法 findViews()

除了通过控件 ID 与控件变量建立连接，本例还需要下列工作。

① 通过临时创建 Calendar 组件得到当前日期，赋值给年月日变量，以使在输入初始日期时以当前日期前 20 年作为进入后的初始状态。

② 监听学历下拉列表，先对字符串列表元素赋值"博士""硕士""学士"，然后将字符串列表加入 ArrayAdapter 中，再将 ArrayAdapter 加入学历选择控件中。

③ 通过 private boolean isValid() {…}方法判断 2 个密码控件输入密码是否一致。只有一致才能具有选择"已阅读并接受网站服务条款"控件（CheckBox）的条件，同时"提交"按钮才能操作。

RegisterActivity.java 代码如下：

```java
package com.easybooks.android.registerpage;
import ...
/**
 * Created by Administrator on 2016/12/8.
 */
public class RegisterActivity extends AppCompatActivity {
    private EditText myName;
    private EditText myPwd;
    private EditText myRePwd;
    private ToggleButton mySex;
    private EditText myBirth;
    private Spinner myDegree;
    private CheckBox myAccept;
    private Button mySubmit;
    private int myYear, myMonth, myDay;
    private String mydegreeTemp;

    @Override
    protected void onCreate(Bundle savedInstanceState) {
        super.onCreate(savedInstanceState);
        setContentView(R.layout.register);
        findViews();
        //1.设置日期事件监听器
        myBirth.setOnClickListener(new View.OnClickListener() {
            @Override
```

```java
            public void onClick(View view) {                    // (1)单击日期框事件
                DatePickerDialog datePickerDialog = new DatePickerDialog(RegisterActivity.this, DateListener,
myYear, myMonth, myDay);                                       // (1)创建 DatePickerDialog 对象，设置初始日期
                datePickerDialog.show();                        // (1)显示 DatePickerDialog 组件（即日期选择对话框）
            }
        });
        //2.②设置学历下拉列表监听器
        Spinner.OnItemSelectedListener listener = new Spinner.OnItemSelectedListener() {
            @Override
            public void onItemSelected(AdapterView<?> parent, View view, int pos, long id) {
                mydegreeTemp = parent.getItemAtPosition(pos).toString();
            }                                        // (2)②将选择项字符串作为值

            @Override
            public void onNothingSelected(AdapterView<?> parent) {
                mydegreeTemp = "未知";                // (2)②没有选择值为未知
            }
        };
        myDegree.setOnItemSelectedListener(listener);
    }

    private void findViews() {
        myName = (EditText) findViewById(R.id.myEditName);
        myPwd = (EditText) findViewById(R.id.myEditPwd);
        myRePwd = (EditText) findViewById(R.id.myEditRePwd);
        mySex = (ToggleButton) findViewById(R.id.myToggleButtonSex);
        myBirth = (EditText) findViewById(R.id.myEditBirth);
        myDegree = (Spinner) findViewById(R.id.mySpinnerDegree);
        myAccept = (CheckBox) findViewById(R.id.myCheckBoxAccept);
        mySubmit = (Button) findViewById(R.id.myButtonSubmit);
        //初始化日历
        Calendar calendar = Calendar.getInstance(Locale.CHINA);     // (2)①创建 Calendar 对象
        Date date = new Date();                                      // (2)①获取当前日期 Date 对象
        calendar.setTime(date);                                      // (2)①为 Calendar 对象设置为当前日期
        myYear = calendar.get(Calendar.YEAR) - 20;                   // (2)①获取 Calendar 对象中的年
        myMonth = calendar.get(Calendar.MONTH);                      // (2)①获取 Calendar 对象中的月
        myDay = calendar.get(Calendar.DAY_OF_MONTH);                 // (2)①获取 Calendar 对象中的日
        myBirth.setText("点击这里选择...");
        //2.②初始化学历
        List<String> list = new ArrayList<String>();                 // (2)②创建字符串数组
        list.add("博士");                                            // (2)②向字符串数组加入元素
        list.add("硕士");
        list.add("学士");
        ArrayAdapter<String> adapter = new ArrayAdapter<String>(this, android.R.layout.simple_spinner_item,
list);                                                              // (2)②创建字符串数组适配器
        adapter.setDropDownViewResource(android.R.layout.simple_spinner_dropdown_item);   //2.②
        myDegree.setAdapter(adapter);    // (2)②将字符串数组适配器与学历控件联系起来
    }
    // (1) 监听日期输入控件单击事件
    private DatePickerDialog.OnDateSetListener DateListener = new DatePickerDialog.OnDateSetListener() {
        @Override
```

```java
        public void onDateSet(DatePicker view, int year, int month, int day) {
            myYear = year;// (2)①当前选择的日期年赋值给全局年变量
            myMonth = month;
            myDay = day;
            //更新日期
            myBirth.setText(myYear + "年" + (myMonth + 1) + "月" + myDay + "日");
        }
    };
    // (2)③myCheckBoxAccept控件单击事件
    public void onCheckBoxClick(View view) {
        if (myAccept.isChecked()) {
            if (isValid(myRePwd)) mySubmit.setEnabled(true);      // (2)③两个密码一致，"提交"按钮可用
            else myAccept.setChecked(false);
        } else mySubmit.setEnabled(false);
    }
    // (2)③判断两个密码是否一致，一致返回true，否则false
    private boolean isValid(EditText editText) {
        String pwd = myPwd.getText().toString();
        String repwd = editText.getText().toString();
        if (!repwd.equals(pwd)) {
            editText.setError("两次输入不一致！");                   // (2)③在密码myRePwd控件中显示错误提示
            return false;
        } else
            return true;
    }
}
```

4. 启动配置

同理，将 activity_main.xml 中的启动页修改为 register.xml：

<include layout="@layout/register" />

在 AndroidManifest.xml 中，修改初始启动的 Activity 为 RegisterActivity（加黑处）：

```xml
<?xml version="1.0" encoding="utf-8"?>
<manifest xmlns:android="http://schemas.android.com/apk/res/android"
    package="com.easybooks.android.registerpage">
    <application
        ...
        <activity
            android:name=".RegisterActivity"
            android:label="@string/app_name"
            ...
        </activity>
    </application>
</manifest>
```

5. 运行效果

运行 APP，输入各项目，结果如图 4.22 所示。

4.2.3 【例三】：图书展示

设计 APP，进行图书展示，效果如图 4.24 所示。

创建图书展示 Android 工程，工程名称为 BookPage。

1. 前期准备

将事先准备好的图书封面图片复制到项目\app\src\main\res\drawable 目录下，如图 4.25 所示。

图 4.24 图书展示界面运行效果　　　　　　图 4.25 准备图书封面图片资源

在当前工程的\res\Values 目录下 strings.xml 文件中，配置字符串文本资源，如下所示。

```
<resources>
    <string name="app_name">新书展示</string>
    <string name="action_settings">Settings</string>

    <string name="title_db0">MySQL 实用教程(第 2 版)</string>
    <string name="info_db0">ISBN：9787121232701</string>
    <string name="title_db1">SQL Server 实用教程(第 4 版)(SQL Server 2012 版)</string>
    <string name="info_db1">ISBN：9787121260384</string>
    <string name="title_db2">SQL Server 实用教程(第 4 版)(SQL Server 2014 版)</string>
    <string name="info_db2">ISBN：9787121266232</string>
    <string name="title_db3">Oracle 实用教程(第 4 版)(Oracle 11g 版)</string>
    <string name="info_db3">ISBN：9787121275722</string>
    <string name="title_db4">Oracle 实用教程(第 4 版)(Oracle 12c 版)</string>
    <string name="info_db4">ISBN：9787121273803</string>

    <string name="title_java0">Java 实用教程(第 3 版)</string>
    <string name="info_java0">ISBN：9787121266225</string>
    <string name="title_java1">Java EE 基础实用教程(第 2 版)</string>
    <string name="info_java1">ISBN：9787121252068</string>
    <string name="title_java2">Java EE 实用教程(第 2 版)</string>
    <string name="info_java2">ISBN：9787121254574</string>

    <string name="title_qt0">Auto-CAD 实用教程(第 4 版)(AutoCAD 2015 中文版)</string>
    <string name="info_qt0">ISBN：9787121260063</string>
    <string name="title_qt1">MATLAB 实用教程(第 4 版)</string>
    <string name="info_qt1">ISBN：9787121291388</string>
```

```
    <string name="title_qt2">Power-Builder 实用教程(第 4 版)</string>
    <string name="info_qt2">ISBN: 9787121210808</string>
    <string name="title_qt3">PHP 实用教程(第 2 版)</string>
    <string name="info_qt3">ISBN: 9787121243394</string>
    <string name="title_qt4">施耐德 PLC 开发及实例(第 2 版)</string>
    <string name="info_qt4">ISBN: 9787121259821</string>
</resources>
```

2. 设计页面

(1) 创建 book.xml 文件,选择"LinearLayout"布局,生成 book.xml 布局文件。

设置线性布局属性:

水平布局: android:orientation="horizontal"

背景颜色: android:background="#aabbcc"

往界面上拖曳一个 ListView 控件,如图 4.26 所示。

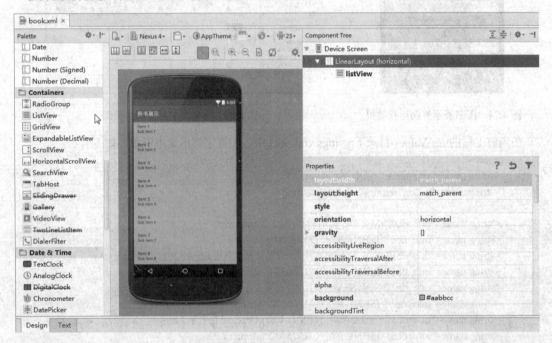

图 4.26　图书列表设计视图

book.xml 文件源码:

```
<?xml version="1.0" encoding="utf-8"?>
<LinearLayout xmlns:android="http://schemas.android.com/apk/res/android"
    android:orientation="horizontal"
    android:layout_width="match_parent"
    android:layout_height="match_parent"
    android:background="#aabbcc">

    <ListView
        android:layout_width="match_parent"
        android:layout_height="wrap_content"
        android:id="@+id/listView" />
</LinearLayout>
```

(2) 创建 listitem.xml 页面，页面布局如下。
① 最外面选择垂直线性布局（L1）。
② 在线性布局（L1）中嵌套水平线性布局（L2）。
③ 在（L2）中加入图片显示控件（id=image）和垂直线性布局（L3）。
④ 在 L3 线性布局中加入 2 个文本框显示控件，id=title，id=info，title 控件显示图书名称，info 控件显示 ISBN。

布局结果如图 4.27 所示。

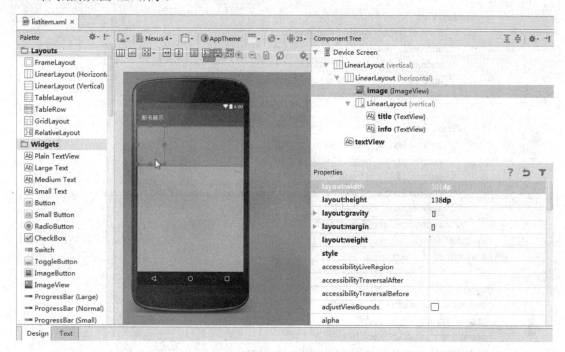

图 4.27　图书项页面设计视图

listitem.xml 文件代码如下：

```xml
<?xml version="1.0" encoding="utf-8"?>
<LinearLayout xmlns:android="http://schemas.android.com/apk/res/android"
    android:orientation="vertical" android:layout_width="match_parent"
    android:layout_height="match_parent"
    android:background="#aabbcc">

    <LinearLayout
        android:orientation="horizontal"
        android:layout_width="match_parent"
        android:layout_height="138dp">

        <ImageView
            android:layout_width="101dp"
            android:layout_height="138dp"
            android:id="@+id/image"
            android:paddingLeft="10dp"
            android:paddingTop="10dp"
            android:background="@android:color/holo_blue_bright" />
```

```xml
                <LinearLayout
                    android:orientation="vertical"
                    android:layout_width="match_parent"
                    android:layout_height="138dp"
                    android:paddingTop="10dp"
                    android:paddingLeft="10dp"
                    android:background="@android:color/holo_blue_bright">

                    <TextView
                        android:layout_width="wrap_content"
                        android:layout_height="wrap_content"
                        android:id="@+id/title"
                        android:textSize="@dimen/abc_text_size_large_material"
                        android:textColor="#000000" />

                    <TextView
                        android:layout_width="wrap_content"
                        android:layout_height="wrap_content"
                        android:id="@+id/info"
                        android:textSize="@dimen/abc_text_size_medium_material"
                        android:textStyle="bold"
                        android:paddingTop="15dp" />
                </LinearLayout>
        </LinearLayout>

        <TextView
            android:layout_width="fill_parent"
            android:layout_height="5dp"
            android:id="@+id/textView"
            android:background="@android:color/holo_blue_light" />

</LinearLayout>
```

3. 功能实现

创建 BookActivity.java 文件。在当前工程→app→java 下选择 com.easybooks.android.bookpage

按右键，选择 New→Java Class，系统显示"Create New Class"对话框：在 Name 处输入"BookActivity"，创建 BookActivity.java 文件。

BookActivity.java 设计思路：

（1）定义 3 个二维数组，分别存放图书封面图片资源、图书名称和 ISBN。二维目的是把它们分成数据库（db）、Java 和其他（qt）三类。

（2）初始 findViews()

① 定义临时 List<Map<String, Object>> list。Map<String, Object>通过字符串与对象之间建立映像。将映像作为 List 的元素。将每一本图书封面图片资源、图书名称和 ISBN 通过 Map 作为 list 项加入其中。

② 定义临时 SimpleAdapter。SimpleAdapter 将加入项内容的临时 list 和显示项内容的 listitem.xml（不写文件扩展名）加入 SimpleAdapter 中，然后将 SimpleAdapter 放入界面的 ListView 控件中：

setAdapter(SimpleAdapter)。

③ 监听界面 ListView 控件的单击事件 setOnItemClickListener(…){ … }

④ 单击事件 onItemClick(…) { … }对应的方法返回的是 ListView 项的位置（序号），需要将位置变成对应的二维数组下标，才能得到对应选择的图书信息。

⑤ 最后用对话框显示其对应的图书名称和 ISBN。

编写 BookActivity.java 文件，代码如下：

```java
package com.easybooks.android.bookpage;
import …
/**
 * Created by Administrator on 2016/12/8.
 */
public class BookActivity extends AppCompatActivity {
    private ListView myListView;
    private int[][] image = {
            {R.drawable.db0, R.drawable.db1, R.drawable.db2, R.drawable.db3, R.drawable.db4},
            {R.drawable.java0, R.drawable.java1, R.drawable.java2},
            {R.drawable.qt0, R.drawable.qt1, R.drawable.qt2, R.drawable.qt3, R.drawable.qt4}
    };              // (1)图书封面图片数组
    private int[][] title = {
            {R.string.title_db0, R.string.title_db1, R.string.title_db2, R.string.title_db3, R.string.title_db4},
            {R.string.title_java0, R.string.title_java1, R.string.title_java2},
            {R.string.title_qt0, R.string.title_qt1, R.string.title_qt2, R.string.title_qt3, R.string.title_qt4}
    };              // (1)书名数组
    private int[][] info = {
            {R.string.info_db0, R.string.info_db1, R.string.info_db2, R.string.info_db3, R.string.info_db4},
            {R.string.info_java0, R.string.info_java1, R.string.info_java2},
            {R.string.info_qt0, R.string.info_qt1, R.string.info_qt2, R.string.info_qt3, R.string.info_qt4}
    };              // (1)书号数组

    @Override
    protected void onCreate(Bundle savedInstanceState) {
        super.onCreate(savedInstanceState);
        setContentView(R.layout.book);
        findViews();
    }

    private void findViews() {
        myListView = (ListView) findViewById(R.id.listView);
        List<Map<String, Object>> list = new ArrayList<Map<String, Object>>();        // (2)①
        for (int i = 0; i < 3; i++) {
            for (int j = 0; j < image[i].length; j++) {
                Map<String, Object> item = new HashMap<String, Object>();
                item.put("image", image[i][j]);
                item.put("title", getResources().getString(title[i][j]));
                item.put("info", getResources().getString(info[i][j]));
                list.add(item);
            }
        }
        SimpleAdapter SimpleAdapter = new SimpleAdapter(this, list, R.layout.listitem, new String[]{"image",
```

```
"title", "info"}, new int[]{R.id.image, R.id.title, R.id.info});                                    // (2)②
            myListView.setAdapter(SimpleAdapter);                                                   // (2)②
            //注册监听器,用消息显示用户点击的图书选项信息
            myListView.setOnItemClickListener(new AdapterView.OnItemClickListener() {               // (2)③
                @Override
                public void onItemClick(AdapterView<?> parent, View view, int position, long id) {  // (2)④
                    String name = "", isbn = "";
                    int row;
                    if (position < 5) {                                 //数据库图书
                        row = 0;
                        name = getString(title[row][position]);
                        isbn = getString(info[row][position]);
                    } else if (position < 8) {                          //Java 图书
                        row = 1;
                        name = getString(title[row][position - 5]);
                        isbn = getString(info[row][position - 5]);
                    } else if (position < 13) {                         //其他图书
                        row = 2;
                        name = getString(title[row][position - 8]);
                        isbn = getString(info[row][position - 8]);
                    }
                    Toast.makeText(BookActivity.this, "您选购了：《" + name + "》，" + isbn, Toast.LENGTH_SHORT).show();
                                                                                                    // (2)⑤
                }
            });
        }
    }
```

4. 启动配置

将 activity_main.xml 中的启动页修改为 book.xml：

`<include layout="@layout/book" />`

将 AndroidManifest.xml 中的启动 Activity 修改为 BookActivity.java：

`<activity android:name=".BookActivity">`

5. 运行效果

运行效果如图 4.24 所示。

第 5 章　Android 多页面与版块

前面介绍的 Android 程序都只有一个页面，本章将介绍多个页面的 Android 应用程序，着重介绍其多个页面之间的跳转、数据传递等交互行为以及 Android 页面的生命周期。还将介绍版块（Fragment）程序设计的基本概念和页面上版块的生命周期。

5.1　Intent 页面间数据传递

5.1.1　Intent 原理

Intent 是 Android 的消息传递机制，主要用于 Android 应用各页面之间或者各项组件之间的数据交换。Intent 负责对应用中一次操作的动作、动作涉及的数据、附加数据进行描述。Android 则根据此 Intent 的描述，负责找到对应的组件，将 Intent 传递给调用的组件，并完成组件的调用。

1. Intent 传递方法

Intent 是一种运行时的绑定机制，它能在程序运行过程中连接两个不同的组件。通过 Intent，程序可以向 Android 表达某种请求或者意愿，Android 会根据意愿的内容选择适当的组件来完成请求。

Android 的三个基本组件（Activity、Service 和 Broadcast Receiver）都是通过 Intent 机制激活的，不同类型的组件有不同的传递 Intent 方式。

（1）激活一个新的 Activity，或者让一个现有的 Activity 做新的操作，可以通过调用 Context.startActivity()或者 Activity.startActivityForResult()方法。

（2）要启动一个新的 Service，或者向一个已有的 Service 传递新的指令，调用 Context.startService()方法或者 Context.bindService()方法，将此方法的上下文对象与 Service 绑定。

（3）Context.sendBroadcast()、Context.sendOrderBroadcast()、Context.sendStickBroadcast()这 3 个方法可以发送 Broadcast Intent。发送之后，所有已注册的并且拥有与之相匹配 IntentFilter 的 BroadcastReceiver 就会被激活。

Intent 一旦发出，Android 都会准确找到相匹配的一个或多个 Activity、Service 或者 Broadcast Receiver 作出响应。所以，不同类型的 Intent 消息不会出现重叠，即 Broadcast 的 Intent 消息只会发送给 Broadcast Receiver，而决不会发送给 Activity 或者 Service。由 startActivity()传递的消息也只会发给 Activity，由 startService()传递的 Intent 只会发送给 Service。

有以下两种形式来使用 Intent。

（1）显式 Intent

显式 Intent 是通过调用 setComponent(ComponentName)或者 setClass(Context,Class)指定了 component 属性的 Intent。通过明确指定具体的组件类，通知应用启动对应的组件。

代码方式可以创建显式 Intent 实例化对象，并设定需要传递的参数信息。由于显式 Intent 指定了具体的组件对象，所以不需要设置 Intent 的其他意图过滤对象。

（2）隐式 Intent

隐式 Intent 是没有指定 Component 属性的 Intent，即没有明确指定组件名，系统根据隐式意图中设置的动作（action）、类别（category）、数据 URI 等来匹配最合适的组件。这些 Intent 需要包含足够的信息，系统根据这些信息，在所有的可用组件中，确定满足此 Intent 的组件。

当一个应用要激活另一个应用中的 Activity 时，只能使用隐式 Intent。根据 Activity 配置的意图过滤器创建一个意图，让意图中的各项参数的值都跟过滤器匹配，这样就可以激活其他应用中的 Activity。所以，隐式 Intent 是在应用与应用之间使用的。

2. Intent 的构成

要在不同的 Activity 之间传递数据，就要在 Intent 中包含相应的内容，一般来说数据中最基本的应该包括以下几方面。

（1）Action

Action 用来指明要实施的动作是什么。一些常用的 Action 如下。

ACTION_CALL activity：启动一个电话。

ACTION_EDIT activity：显示用户编辑的数据。

ACTION_MAIN activity：作为 Task 中第一个 Activity 启动。

ACTION_SYNC activity：同步手机与数据服务器上的数据。

ACTION_BATTERY_LOW broadcast receiver：电池电量过低警告。

ACTION_HEADSET_PLUG broadcast receiver：插拔耳机警告。

ACTION_SCREEN_ON broadcast receiver：屏幕变亮警告。

ACTION_TIMEZONE_CHANGED broadcast receiver：改变时区警告。

（2）Data

Data 是执行动作要操作的数据，Android 采用指向数据的一个 URI 来表示。

例如，在联系人应用中，一个指向某联系人的 URI 可能为"content://contacts/1"。这种 URI 表示，通过 ContentURI 这个类来描述。以下是一些 action/data 对及其要表达的意图。

VIEW_ACTION content://contacts/1：显示标识符为"1"的联系人的详细信息。

EDIT_ACTION content://contacts/1：编辑标识符为"1"的联系人的详细信息。

VIEW_ACTION content://contacts/：显示所有联系人的列表。

PICK_ACTION content://contacts/：显示所有联系人的列表，并且允许用户在列表中选择一个联系人，然后把这个联系人返回给父 activity。例如，电子邮件客户端可以使用这个 Intent，要求用户在联系人列表中选择一个联系人。

（3）Category

Category 表示一个类别字符串，包含了有关处理该 Intent 组件的种类信息。例如：

LAUNCHER_CATEGORY：Intent 的接受者应该在 Launcher 中作为项级应用出现。

ALTERNATIVE_CATEGORY：当前的 Intent 是一系列的可选动作中的一个，这些动作可以在同一块数据上执行。

一个 Intent 对象可以有任意个 category，Intent 类定义了许多 category 方法。

addCategory()：为一个 Intent 对象增加一个 category。

removeCategory()：用于删除一个 category。

getCategories()：用于获取 Intent 所有的 category。

（4）Type

Type 用于显式指定 Intent 的数据类型 MIME。例如，一个组件是可以显示图片数据而不能播放声音文件。一般 Intent 的数据类型能够根据数据本身进行判定，很多情况下，data 类型可在 URI 中找到，例如 content:开头的 URI，表明数据由设备上的 content provider 提供。但是通过设置这个属性，可以强制采用显式指定的类型而不再进行推导。

（5）component

component 用于指定 Intent 的目标组件的类名称。通常 Android 会根据 Intent 中包含的其他属性的信息，例如 action、data/type、category 进行查找，最终找到一个与之匹配的目标组件。但是，如果指定了 component 的属性，将直接使用它指定的组件，而不再执行上述查找过程。指定了这个属性以后，Intent 的其他所有属性都是可选的。

（6）extras

extras 是其他所有附加信息的集合，使用 extras 可以为组件提供扩展信息。例如，如果要执行"发送电子邮件"这个动作，可以将电子邮件的标题、正文等保存在 extras 里，传给电子邮件发送组件。附加信息可以使用 putExtras()和 getExtras()作为 Bundle 来读和写。

总之，action、data/type、category 和 extras 一起形成了一种语言，这种语言使系统能够理解诸如"查看某联系人的详细信息"之类的短语。随着应用不断加入到系统中，它们可以添加新的 action、data/type、category 来扩展这种语言。应用也可以提供自己的 Activity 来处理已经存在的这样的"短语"，从而改变这些"短语"的行为。

3. Intent 解析机制

Intent 解析机制主要是通过查找已注册在 AndroidManifest.xml 中的所有<intent-filter>及其中定义的 Intent，最终找到匹配的 Intent。在这个解析过程中，Android 是通过 Intent 的 action、type、category 这三个属性来进行判断的，判断方法如下。

（1）如果 Intent 指明定了 action，则目标组件的 IntentFilter 的 action 列表中就必须包含这个 action，否则不能匹配。

（2）如果 Intent 没有提供 type，系统将从 data 中得到数据类型。和 action 一样，目标组件的数据类型列表中必须包含 Intent 的数据类型，否则不能匹配。

（3）如果 Intent 中的数据不是 content:类型的 URI，而且 Intent 也没有明确指定 type，将根据 Intent 中数据的 scheme（例如 http:或者 mailto:）进行匹配。Intent 的 scheme 也必须出现在目标组件的 scheme 列表中。

（4）如果 Intent 指定了一个或多个 category，这些类别必须全部出现在组件的类别列表中。例如 Intent 中包含了两个类别：LAUNCHER_CATEGORY 和 ALTERNATIVE_CATEGORY，解析得到的目标组件必须至少包含这两个类别。

4. Activity 的 Intent 数据传递

（1）Activity 间的数据传递

一个 Android 应用程序可能包含多个 Activity，要从一个 Activity 切换到另一个 Activity，必须通过 Intent，因为 Intent 存储着切换时所需的重要信息。如图 5.1 所示为 Activity01 通过 Intent 切换到 Activity02。

如果想要在两个 Activity 切换时附带额外数据，可以将该项数据存储在 Bundle 内。Bundle 依附在 Intent 上，是一个专门用来存储附加数据的对象。如图 5.2 所示。

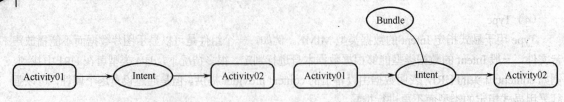

图 5.1 Activity01 通过 Intent 切换到 Activity02　　图 5.2 Bundle 的作用

具体包括以下几方面。

① 直接向 Intent 对象中传入键值对，相当于 Intent 对象具有 Map 键值对功能。

② 定义一个 Bundle 对象，在该对象中加入键值对，然后将该对象加入 Intent 中。

③ 向 Intent 中添加 ArrayList 集合对象。

④ Intent 传递 Object 对象，被传递对象实现 Parcelable 接口，或者实现 Serializable 接口。

（2）Activity 退出时的返回结果

① 通过 startActivityForResult 方式启动一个 Activity。

② 新 activity 设定 setResult 方法，通过该方法可以传递 responseCode 和 Intent 对象。

③ 在 MainActivity 中覆写 onActivityResult 方法，新 activity 一旦退出，就会执行该方法。

5.1.2　基本数据类型传递方式（【例一】：登录响应）

一个 Activity 可以通过 Bundle 传递数据到下一个 Activity，而基本数据类型与其对应的数组类型和字符串（其实就是字符数组）都是经常传递的数据类型。如图 5.3 所示。

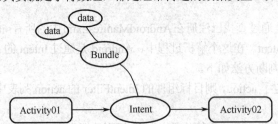

图 5.3　Activity 通过 Bundle 传递数据

【例一】在第 4 章【例一】的基础上实现，应用 Intent 将用户登录界面输入的信息传递到登录响应页面。输入用户名和密码，如图 5.4（a）所示。

单击"登录"按钮，如果密码正确，系统显示登录响应页面，如图 5.4（b）所示。单击"返回"按钮，系统重新显示如图 5.4（a）所示。

(a)　　　　　　　　　(b)

图 5.4　登录和登录响应

如果密码不正确，系统显示登录响应页面，如图 5.5（a）所示。单击"返回"按钮，系统重新显示如图 5.5（b）所示。

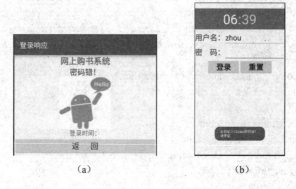

图 5.5 登录响应

1. 设计页面

（1）打开 Android 工程 LoginPage，添加设计一个登录响应页面 welcome.xml，如图 5.6 所示。

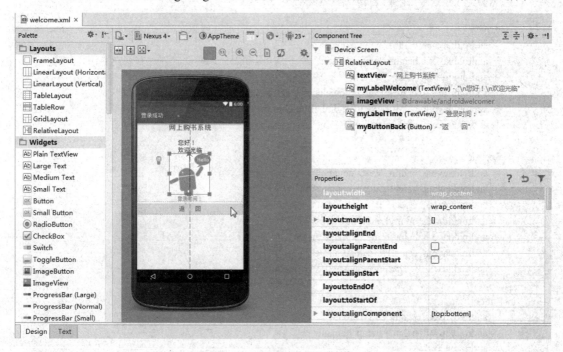

图 5.6 设计视图

将图片资源 androidwelcomer.gif 预先复制到项目 app\src\main\res\drawable 目录下。

welcome.xml 代码如下：

```xml
<?xml version="1.0" encoding="utf-8"?>
<RelativeLayout xmlns:android="http://schemas.android.com/apk/res/android"
    android:layout_width="match_parent" android:layout_height="match_parent">

    <TextView
        android:layout_width="wrap_content"
```

```xml
        android:layout_height="wrap_content"
        android:text="网上购书系统"
        android:id="@+id/textView"
        android:layout_alignParentTop="true"
        android:layout_centerHorizontal="true"
        android:textSize="@dimen/abc_text_size_headline_material"
        android:textStyle="bold"
        android:textColor="@android:color/holo_blue_dark" />

    <TextView
        android:layout_width="wrap_content"
        android:layout_height="wrap_content"
        android:textAppearance="?android:attr/textAppearanceLarge"
        android:text="\n 您好！\n 欢迎光临"
        android:id="@+id/myLabelWelcome"
        android:layout_below="@+id/textView"
        android:layout_centerHorizontal="true"
        android:textSize="@dimen/abc_text_size_large_material"
        android:textColor="@android:color/holo_red_dark" />

    <ImageView
        android:layout_width="wrap_content"
        android:layout_height="wrap_content"
        android:id="@+id/imageView"
        android:src="@drawable/androidwelcomer"
        android:layout_below="@+id/myLabelWelcome"
        android:layout_centerHorizontal="true" />

    <TextView
        android:layout_width="wrap_content"
        android:layout_height="wrap_content"
        android:text="登录时间："
        android:id="@+id/myLabelTime"
        android:layout_below="@+id/imageView"
        android:layout_centerHorizontal="true"
        android:textSize="@dimen/abc_text_size_medium_material" />

    <Button
        android:layout_width="fill_parent"
        android:layout_height="wrap_content"
        android:text="返    回"
        android:id="@+id/myButtonBack"
        android:layout_below="@+id/myLabelTime"
        android:layout_centerHorizontal="true"
        android:textSize="@dimen/abc_text_size_large_material"
        android:onClick="onBackClick" />                    //单击"返回"按钮，执行该方法
</RelativeLayout>
```

（2）修改 strings.xml 文件，代码如下：

```xml
<string name="app_name">登录响应</string>
```

在 Androidmanifest.xml 文件中引用这个字符串作为 APP 的标题。

```
<application
    android:label="@string/app_name"
    ...
</application>
```

2. 功能实现

前面我们已经介绍，系统默认启动的页面在 activity_main.xml 文件中定义的。在 activity_main.xml 中设置下列内容：

```
<include layout="@layout/login" />
```

系统启动时显示 login.xml 定义的界面，当时我们没有过多地关注 java 类，也没有关注多个页面之间的转换。

现在我们需要重点关注在多个页面之间的转换及其数据传递。

具体过程和关系说明如下。

（1）APP 启动页面 Activity 配置

① 在 AndroidManifest.xml 中，将新添加的欢迎页面的 Activity（WelcomeActivity）配置其中，同时将登录页面的 Activity（LoginActivity）作为启动项。

```
<application
    android:allowBackup="true"
    android:icon="@mipmap/ic_launcher"
    android:label="@string/app_name"
    android:supportsRtl="true"
    android:theme="@style/AppTheme">
    <activity
        android:name=".LoginActivity"
        android:label="@string/app_name"
        android:theme="@style/AppTheme.NoActionBar">
        <intent-filter>
            <action android:name="android.intent.action.MAIN" />
            <category android:name="android.intent.category.LAUNCHER" />
        </intent-filter>
    </activity>
    <activity android:name=".WelcomeActivity"></activity>
</application>
```

其中 LoginActivity 中 `<action android:name="android.intent.action.MAIN" />` 表示 LoginActivity 为启动页面。

② 在 LoginActivity.java 文件中的 onCreate() 中指定 login.xml 作为页面。

```
protected void onCreate(Bundle savedInstanceState) {
    super.onCreate(savedInstanceState);
    setContentView(R.layout.login);
    findViews();
}
```

（2）在 Login.xml 定义的页面中"登录"按钮如下：

```
<Button
    android:layout_width="wrap_content"
    android:layout_height="wrap_content"
    android:text="登录"
    android:id="@+id/myButtonOk"
    android:textSize="@dimen/abc_text_size_display_1_material"
```

```
android:textStyle="bold"
android:layout_weight="0.25"
android:onClick="onLoginClick" />
```

单击"登录"按钮,执行 LoginActivity.java 文件中的 onLoginClick 事件代码,实现下列功能。

① 获得登录界面交互信息。

② 定义数据传递机制:定义 Intent 数据传递机制引用 intent,用于传输到 WelcomeActivity.class 指定的页面,同时定义 Bundle 数据绑定引用 bundle。

③ 将获得的登录界面交互信息加入 bundle 绑定,前面的参数为数据名称,后面的参数为数据内容。

④ bundle 绑定放入 intent 传输机制。

⑤ 启动页面数据传输。

(3) 当重新返回当前(Login.xml)页面时,onActivityResult()方法被自动执行。

LoginActivity.java 文件代码如下:

```java
package com.easybooks.android.loginpage;
import …
/**
 * Created by Administrator on 2016/10/26.
 */
public class LoginActivity extends AppCompatActivity {
    private EditText myName;
    private EditText myPwd;
    private Button myOk;
    private TimePicker myTime;
    private DatePicker myDate;

    @Override
    protected void onCreate(Bundle savedInstanceState) {
        super.onCreate(savedInstanceState);
        setContentView(R.layout.login);
        findViews();
    }

    private void findViews() {
        myName = (EditText) findViewById(R.id.myTextName);
        myPwd = (EditText) findViewById(R.id.myTextPwd);
        myOk = (Button) findViewById(R.id.myButtonOk);
        myTime = (TimePicker) findViewById(R.id.myTimePicker);
        myDate = (DatePicker) findViewById(R.id.myDatePicker);
        myTime.setIs24HourView(true);              //时间采用 24 小时制
    }

    public void onLoginClick(View view) {
        String name = myName.getText().toString();     // (2)①获得登录名
        String pass = myPwd.getText().toString();      // (2)①获得密码
        int year, month, day, hour, minute;
        year = myDate.getYear();                       // (2)①获得日期中的年
        month = myDate.getMonth() + 1;                 // (2)①获得日期中的月
        day = myDate.getDayOfMonth();                  // (2)①获得日期中的日
```

```java
            hour = myTime.getCurrentHour();                    // (2)①获得日期中的时
            minute = myTime.getCurrentMinute();                // (2)①获得日期中的分
            Intent intent = new Intent(this, WelcomeActivity.class);   //(2)②定义 intent 数据传递机制
            Bundle bundle = new Bundle();                      // (2)②定义 bundle 数据传递绑定
            bundle.putString("name", name);                    // (2)③字符串 name 内容绑定到 bundle
            bundle.putString("pass", pass);                    // (2)③
            bundle.putInt("year", year);                       // (2)③
            bundle.putInt("month", month);                     // (2)③
            bundle.putInt("day", day);                         // (2)③
            bundle.putInt("hour", hour);                       // (2)③
            bundle.putInt("minute", minute);                   // (2)③
            intent.putExtras(bundle);                          // (2)④bundle 放入 intent 传输机制
            //startActivity(intent);                           // (2)④不要求新 Activity 退出时返回值
            startActivityForResult(intent, 200);               // (2)⑤启动数据传输,请求码 200
        }

        protected void onActivityResult(int requestCode, int resultCode, Intent data) {       // (3)
            if (resultCode == 101) Toast.makeText(this, "重新登录", Toast.LENGTH_LONG).show();
            else if (resultCode == 404) {
                Toast.makeText(this, "此前输入" + data.getStringExtra("pass") + "密码错! \n 请重输", Toast.LENGTH_LONG).show();
                myPwd.setText("");                             //清空密码框
            }
        }
    }
```

当重新调用本页面时,onActivityResult(int requestCode, int resultCode, Intent data)方法被自动执行。其中:

  requestCode:带回请求码。

  resultCode:带回响应码。

  data:带回数据。

(4)创建 WelcomeActivity.java 文件,代码如下:

```java
package com.easybooks.android.loginpage;
import …
/**
 * Created by Administrator on 2016/10/26.
 */
public class WelcomeActivity extends AppCompatActivity {
    private TextView myWelcome;
    private TextView myTime;
    private String name, pass;
    private int resultCode;                         //响应码:登录成功 101; 错误 404

    @Override
    protected void onCreate(Bundle savedInstanceState) {
        super.onCreate(savedInstanceState);
        setContentView(R.layout.welcome);           //设置显示页面为 welcome.xml
        findViews();
        showWelcome();
    }
```

```java
    private void findViews() {
        myWelcome = (TextView) findViewById(R.id.myLabelWelcome);
        myTime = (TextView) findViewById(R.id.myLabelTime);
    }

    private void showWelcome() {
        Bundle bundle = getIntent().getExtras();          //定义 Bundle 传递数据对象 bundle
        name = bundle.getString("name");                  //从 bundle 中得到 name 内容（输入的用户名）
        pass = bundle.getString("pass");                  //从 bundle 中得到 pass 内容（输入的密码）
        if (pass.equals("123456")) {
            myWelcome.setText("\n" + name + " 您好！\n     欢迎光临");
            int year, month, day, hour, minute;
            year = bundle.getInt("year");
            month = bundle.getInt("month");
            day = bundle.getInt("day");
            hour = bundle.getInt("hour");
            minute = bundle.getInt("minute");
            myTime.setText("登录时间： " + year + "-" + month + "-" + day + " " + hour + ":" + minute);
            resultCode = 101;
        } else {
            myWelcome.setText("密码错！ ");
            resultCode = 404;
        }
        Intent data = new Intent(this, MainActivity.class);
        data.putExtras(bundle);
        setResult(resultCode, data);
    }

    public void onBackClick(View view) {
        finish();                                         //结束当前 Activity，回到前面一个 Activity
    }
}
```

5.1.3 对象数据类型传递方式（【例二】：注册成功直接登录）

Bundle 除了可以存储基本数据类型以外还可以存储对象，如图 5.7 所示，但必须是 Serializable 对象或 Parcelable 对象。

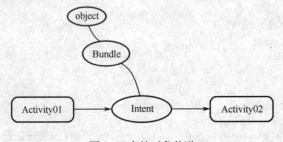

图 5.7 存储对象传递

使用 Serializable 对象比较简单，因为对象的解组（unmarshalling/marshalling）是交给执行环境处理的,不需要自行编写程序。而 Parcelable 对象需要自行编写程序来实现 Parcelable 与 Parcelable.Creator 接口以达到解组功能,虽然比较麻烦但执行效率较好。

【例二】在第 4 章【例二】的基础上实现,将注册用户信息封装在 User 对象中传递到成功页面,注册成功后可直接登录,再将当前用户登录的时间日期封装在 Time 对象中传递到欢迎页面。

1. 编写 Java 对象

(1) 编写存储用户信息的 User 对象,包含登录名、密码、性别、出生时间、学历等属性。除了构造方法,设置和获得属性值方法,还有将所有属性值连接成一个字符串的方法。

User.java 代码如下:

```java
package com.easybooks.android.registerpage;
import java.io.Serializable;
/**
 * Created by Administrator on 2016/10/28.
 */
public class User implements Serializable{
    private String name;            //登录名
    private String pwd;             //密码
    private int sex;                //性别
    private String birth;           //出生时间
    private String degree;          //学历

    public User(String name,String pwd,int sex,String birth,String degree) {
        this.name = name;
        this.pwd = pwd;
        this.sex = sex;
        this.birth = birth;
        this.degree = degree;
    }

    @Override
    public String toString() {      //将所有属性值连接成一个字符串
        String userInfo = name + " " + ((sex == 1) ? "先生" : "女士") + ", 恭喜您注册成功!\n 您的注册信息为:\n 出生日期      " + birth + "\n 学       历       " + degree;
        return userInfo;
    }
    public String getName(){
        return name;
    }
    public void setName(String name) {
        this.name = name;
    }
    public String getPwd(){
        return pwd;
    }
    public void setPwd(String pwd) {
        this.pwd = pwd;
    }
    public int getSex(){
        return sex;
```

```java
    }
    public void setSex(int sex) {
        this.sex = sex;
    }
    public String getBirth(){
        return birth;
    }
    public void setBirth(String birth) {
        this.birth = birth;
    }
    public String getDegree(){
        return degree;
    }
    public void setDegree(String degree) {
        this.degree = degree;
    }
}
```

（2）编写存储时间日期的 Time 对象，包含年、月、日、时、分等属性。除了构造方法，设置和获得属性值方法，还有将所有属性值连接成一个字符串的方法。

Time.java 代码如下：

```java
package com.easybooks.android.registerpage;
import java.io.Serializable;
/**
 * Created by Administrator on 2016/11/3.
 */
public class Time implements Serializable {
    private int year;           //年
    private int month;          //月
    private int day;            //日
    private int hour;           //时
    private int minute;         //分

    public Time(int year, int month, int day, int hour, int minute) {
        this.year = year;
        this.month = month;
        this.day = day;
        this.hour = hour;
        this.minute = minute;
    }

    @Override
    public String toString() {          //将所有属性值连接成一个字符串
        String timeInfo = "登录时间：" + year + "-" + month + "-" + day + " " + hour + ":" + minute;
        return timeInfo;
    }
    public int getYear() {
        return year;
    }
    public void setYear(int year) {
        this.year = year;
```

```
    }
    public int getMonth() {
        return month;
    }
    public void setMonth(int month) {
        this.month = month;
    }
    public int getDay() {
        return day;
    }
    public void setDay(int day) {
        this.day = day;
    }
    public int getHour() {
        return hour;
    }
    public void setHour(int hour) {
        this.hour = hour;
    }
    public int getMinute() {
        return minute;
    }
    public void setMinute(int minute) {
        this.minute = minute;
    }
}
```

2. 设计页面

(1) 打开 Android 工程 RegisterPage,添加设计一个注册成功页面 success.xml,如图 5.8 所示。

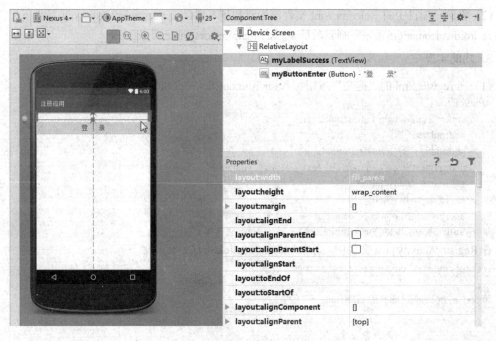

图 5.8 注册成功页面设计视图

success.xml 文件代码如下:

```xml
<?xml version="1.0" encoding="utf-8"?>
<RelativeLayout xmlns:android="http://schemas.android.com/apk/res/android"
    android:layout_width="match_parent" android:layout_height="match_parent"
    android:background="@android:color/background_light">

    <TextView
        android:layout_width="fill_parent"
        android:layout_height="wrap_content"
        android:id="@+id/myLabelSuccess"
        android:layout_alignParentTop="true"
        android:layout_centerHorizontal="true"
        android:background="@android:color/background_light"
        android:textSize="@dimen/abc_text_size_large_material"
        android:textColor="@color/colorAccent"
        android:textStyle="bold" />

    <Button
        android:layout_width="fill_parent"
        android:layout_height="wrap_content"
        android:id="@+id/myButtonEnter"
        android:layout_below="@+id/myLabelSuccess"
        android:layout_centerHorizontal="true"
        android:text="登    录"
        android:textSize="@dimen/abc_text_size_large_material"
        android:onClick="onEnterClick" />

</RelativeLayout>
```

其中:TextView 控件用于显示信息,单击"登录"按钮,执行 onEnterClick 方法。

(2)将例一项目中的 welcome.xml 文件添加复制到本项目 app\src\main\res\layout 目录下。将图片资源 androidwelcomer.gif 预先复制到项目 app\src\main\res\drawable 目录下。

3. 功能实现

(1)在 register.xml 的"提交"按钮加入 android:onClick="onSubmitClick"属性:

```xml
<Button
    style="?android:attr/buttonStyleSmall"
    android:text="提    交"
    android:id="@+id/myButtonSubmit"
    android:textSize="@dimen/abc_text_size_large_material"
    android:enabled="false"
    android:gravity="left"
    android:textAlignment="center"
    android:onClick="onSubmitClick"  />
```

在 RegisterActivity.java 文件中,添加 onSubmitClick 方法,代码如下:

```java
package com.easybooks.android.registerpage;

import android.app.DatePickerDialog;
...

/**
```

```
 * Created by Administrator on 2016/10/27.
 */
public class RegisterActivity extends AppCompatActivity {
    ...
    private boolean isValid(EditText editText) {  ...  }

    public void onSubmitClick(View view) {
        String name = myName.getText().toString();                              //①
        String pwd = myPwd.getText().toString();                                //①
        int sex = Integer.parseInt(mySex.getText().toString().equals("男") ? "1" : "0");
        String birth = myBirth.getText().toString();                            //①
        String degree = mydegreeTemp;                                           //①
        User user = new User(name, pwd, sex, birth, degree);                    //②
        Intent intent = new Intent(this, SuccessActivity.class);                //③
        Bundle bundle = new Bundle();                                           //③
        bundle.putSerializable("user", user);                                   //④
        intent.putExtras(bundle);                                               //⑤
        startActivity(intent);                                                  //⑥
    }
}
```

其中：

① 获得注册界面交互信息。

② 定义 User 对象实例 user,将用户界面各项作为 user 的属性。

③ 定义数据传递机制：定义 Intent 数据传递机制引用 intent，用于传输到 SuccessActivity.class 指定的页面，同时定义 Bundle 数据绑定引用 bundle。

④ 将 User 对象实例 user 加入 bundle 绑定。

⑤ bundle 绑定放入 intent 传输机制。

⑥ 启动页面数据传输。

（2）创建 SuccessActivity.java 文件，显示 success.xml 页面。功能分以下几个方面。

① 在页面的文本框中先通过 User 对象显示注册用户信息。因为进入前面页面的 onSubmitClick() 方法中已经将注册用户信息放入 User 对象 user，并将其放入接口进行传递，这里获取传递的 user，通过 user 的 toString()方法得到注册用户信息字符串。

② 用户单击本页面"登录"按钮，执行 onEnterClick()方法。

③ 从接口的 user 中获得注册用户名。

④ 将当前时间放入 Time 对象 time。

⑤ 将注册用户名作为登录名和 Time 对象 time 传递到 Welcome.xml 页面。

代码如下：

```
package com.easybooks.android.registerpage;
import android.content.Intent;
...
/**
 * Created by Administrator on 2016/10/28.
 */
public class SuccessActivity extends AppCompatActivity {
    private TextView mySuccess;
    private Button myEnter;
```

```java
    @Override
    protected void onCreate(Bundle savedInstanceState) {
        super.onCreate(savedInstanceState);
        setContentView(R.layout.success);                    //显示 success.xml 页面
        findViews();
        showSuccess();            //在页面文本框中显示 User 对象信息（注册页面交互内容）
    }

    private void findViews() {
        mySuccess = (TextView) findViewById(R.id.myLabelSuccess);
        myEnter = (Button) findViewById(R.id.myButtonEnter);
    }

    private void showSuccess() {
        Bundle bundle = getIntent().getExtras();             //(2)①获取接口信息
        Object user = bundle.getSerializable("user");        //(2)①得到 user 实例
        mySuccess.setText(user.toString());                  //(2)①mySuccess 文本框显示注册信息
    }

    public void onEnterClick(View view) {                    //(2)②
        Bundle bundle = getIntent().getExtras();             //(2)③获取接口信息
        User user = (User) bundle.getSerializable("user");   //(2)③得到 user 实例
        String name = user.getName();                        //(2)③得到 user 实例的登录名

        int year, month, day, hour, minute;
        Calendar calendar = Calendar.getInstance();          //(2)④
        year = calendar.get(Calendar.YEAR);                  //(2)④
        month = calendar.get(Calendar.MONTH) + 1;            //(2)④
        day = calendar.get(Calendar.DAY_OF_MONTH);           //(2)④
        hour = calendar.get(Calendar.HOUR_OF_DAY);           //(2)④
        minute = calendar.get(Calendar.MINUTE);              //(2)④
        Time time = new Time(year, month, day, hour, minute); //(2)④

        Intent intent = new Intent(this, WelcomeActivity.class); //(2)⑤
        bundle.putString("name", name);                      //(2)⑤
        bundle.putSerializable("time", time);                //(2)⑤
        intent.putExtras(bundle);                            //(2)⑤
        startActivity(intent);                               //(2)⑤
    }
}
```

（3）将例 4 项目中的 WelcomeActivity.java 文件添加复制到本项目 app\src\main\java\com\easybooks\android\registerpage 目录下。

修改命名空间如下：

```
package com.easybooks.android.registerpage;
```

代替 showWelcome()方法，代码如下（特别注意加黑代码）：

```java
private void showWelcome() {
    Bundle bundle = getIntent().getExtras();
    String name = bundle.getString("name");
    myWelcome.setText("\n" + name + " 您好！\n       欢迎光临");
```

```
        Object time = bundle.getSerializable("time");
        myTime.setText(time.toString());
}
```

4. Activity 配置

（1）在 AndroidManifest.xml 中，将新添加的注册成功页面（SuccessActivity）和欢迎页面（WelcomeActivity）配置其中（加黑处）：

```xml
<?xml version="1.0" encoding="utf-8"?>
<manifest xmlns:android="http://schemas.android.com/apk/res/android"
    package="com.easybooks.android.registerpage">

    <application
        android:allowBackup="true"
        android:icon="@mipmap/ic_launcher"
        android:label="注册应用"
        android:supportsRtl="true"
        android:theme="@style/AppTheme">
        <activity
            android:name=".RegisterActivity"
            android:theme="@style/AppTheme.NoActionBar">
            <intent-filter>
                <action android:name="android.intent.action.MAIN" />

                <category android:name="android.intent.category.LAUNCHER" />
            </intent-filter>
        </activity>
        <activity
            android:name=".SuccessActivity"
            android:label="@string/register_success"/>
        <activity
            android:name=".WelcomeActivity"
            android:label="@string/login_success"/>
    </application>
</manifest>
```

（2）在项目 strings.xml 中配置文本字符串资源，如下：

```xml
<resources>
    <string name="register_success">注册成功</string>
    <string name="login_success">登录成功</string>
    <string name="action_settings">Settings</string>
</resources>
```

5. 运行效果

运行 APP，系统显示注册页面，输入信息，如图 5.9（a）所示。单击"提交"按钮，系统显示如图 5.9（b）所示。此时单击"登录"按钮，系统显示如图 5.9（c）所示。

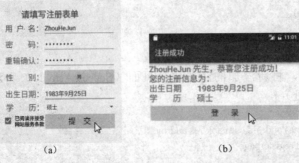

图 5.9 注册功能运行效果

5.2 Activity 生命周期

5.2.1 Activity 概述

Activity 是 Android 组件中最基本也是最为常用的四大组件（Activity、Service、ContentProvider、BroadcastReceiver）之一。Activity 中所有操作都与用户密切相关，是一个负责与用户交互的组件，可以通过 setContentView(View) 显示指定页面，可以监听并处理用户的事件做出响应，在 Activity 之间通过 Intent 进行通信。

一个应用程序通常有多个页面，包含多个 Activity，AndroidManifest.xml 文件中需要加入一个应用所有的 Activity，但包含

<action android:name="android.intent.action.MAIN" />

项的 Activity 为启动页面。

Activity 可以启动其他的 Activity 进行相关的操作。当启动其他的 Activity 时，当前的 Activity 将会停止，新的 Activity 将会压入栈中，同时获取用户焦点，这时就可在新的 Activity 上操作了。因为栈是先进后出的原则，当用户按【Back】键时，当前的这个 Activity 销毁，前一个 Activity 重新恢复。

5.2.2 生命周期的基本概念

如果以 MVC(Model-View-Controller)设计模式来说，Android 应用程序的一页其实是由 layout 文件（View）与 Activity 类（Controller）组成的，Activity 扮演着控制页面流程的角色，所以说是一个页面最核心的部分。

Activity 控制的页面从产生到结束，会经历 7 个阶段，这 7 个阶段就是一个页面的生命周期。Android 为了方便开发者能够轻易指定每个阶段要执行什么程序，而创建了 Activity 类，并在其中定义了与生命周期有关的 7 种方法，如表 5.1 所示。开发者改写这些方法，便于加入想要执行的程序内容。

表 5.1 与生命周期有关的方式

方法	说明	可否删除程序	下一个阶段
onCreate()	当 Activity 第一次被创建时调用，用来加载必要的数据到内存中，完成初始化，方便后续使用。包含： 载入 UI 画面：例如调用 setContentView()以载入 layout 文件内容 初始化 UI 组件：例如调用 findViewById()以取得对应的 UI 组件	否	onStart()
onStart()	当 Activity 画面即将显示前会调用	否	onResume()

续表

方　法	说　明	可否删除程序	下一个阶段
onResume()	当每次 Activity 画面要显示时调用。如果画面显示时所要使用的特定功能很耗电，不想太早打开，可以放在这里打开。例如打开 GPS 定位功能	否	onPause()
onPause()	当前的 Activity 画面无法完全显示时（例如半遮状态）调用。应该在此阶段释放较耗电的资源。例如停止 GPS 定位功能	是	当 Activity 画面完全被取代会调用 onStop()。画面在半遮状态恢复时会直接回到 onResume()，而不会经过 onRestart()
onStop()	当前 Activity 画面被其他 Activity 画面完全取代调用	是	如果 Activity 要结束会调用 onDestroy()；如果要恢复此 Activity 到可以显示画面，会先调用 onRestart()
onRestart()	当 Activity 从 onStop()状态要恢复到 onStart()状态时调用	否	onStart()
onDestroy()	Activity 要结束之前调用。建议此阶段释放所有尚未释放的资源	是	

说明：

（1）上述每一个方法都代表 Activity 生命周期的一个阶段。开发者可以自行创建一个类继承 Activity 类，并且按照自己的需要而改写对应的方法。Activity 执行到指定的阶段，就会自动调用被改写的对应方法。

改写这 7 个方法都必须调用父类对应被改写的方法，例如改写 onStart()时必须加上 super.onStart()，否则会弹出 Exception。

（2）Activity 的执行程序可否被强制停止并删除。一般而言，Activity 页面从加载到显示阶段 (onCreate()→onStart()→onResume()→页面显示）都是重要阶段，因为此时用户正等待画面打开，如果程序被删除，画面就无法显示，用户就会不满意。Activity 页面如果处于休眠甚至结束阶段（onPause() →onStop→onDestroy()→页面结束），这些阶段不仅程序已经停止，就连画面也没有显示，即使强制删除 Activity 程序，对用户影响也不大。

（3）现在的智能手机，大多具有多任务（Multi-Task）功能，例如，使用手机听音乐的同时还可以运行其他多个应用程序。这种多任务功能虽然方便，但是每多运行一个应用程序，就必定多耗费系统的内存。内存是有限的，所以当同时运行的程序越多，系统整体运行就会越慢，甚至不稳定。

为了兼顾应用程序的正常运行与内存的有效利用，Android 有自己的一套程序管理模式，其中最重要的部分就是前面所述 Activity 生命周期的管理。按照表 5.1 可以绘制出如图 5.10 所示的 Activity 的生命周期。

（4）Activity 的 7 种方法决定了一个 Activity 的完整生命周期，但是 Activity 不一定会运行所有方法，而有些方法不一定只运行一次。一个 Activity 会经历哪些方法，大多是由 Android 系统按照用户操作与系统资源使用情况加以控制。将 Activity 经历过的方法结合起来，其实代表的就是一个完整的程序流程，也可以称作进程（process）。

进程由 Android 系统掌控，只要内存不足时，Android 系统可能会随时按照进程的重要程度决定要终止的进程。最不重要的进程，最容易被终止。从重要到不重要的顺序排列如下。

- 前台进程（foreground process）：用户正在使用的进程，而且该进程的画面正显示在屏幕上。这种进程几乎不会被终止，除非现有内存空间少得可怜，而且没有其他不重要的进程可以终止，前台进程才可能被终止。

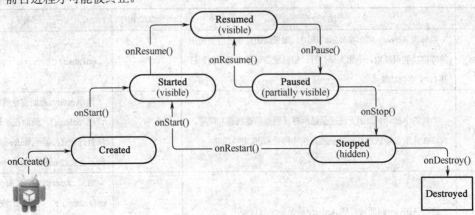

图 5.10 Activity 的生命周期

Activity 的 onCreate()、onStart()、onResume()方法被调用时，该 Activity 就会进入到前台进程。
- 可视进程（visible process）：虽然不是前台进程，但用户仍然可以看到该进程所显示的画面。例如按电源键会开启键盘锁，此时仍可看到主画面的背景图，解开键盘锁后仍然会回到该进程。进入到这个进程，Activity 的 onPause()方法会被调用。
- 服务进程（service process）：服务进程就是 Android 的 Service 功能，与前两项进程属于 Activity 有所不同，该进程启动后会保持运行状态，以持续对用户提供服务，例如播放 MP3 音乐或从网络下载数据。
- 后台进程（background process）：此进程的画面既没有显示，对用户也没有直接影响，进入到这个进程，Activity 的 onStop()方法会被调用。
- 空进程（empty process）：当后台进程被终止，会将所占的内存空间释放，该进程就会由后台进程进入到空进程，但 Activity 仍然存在（只要 Activity 的 onDestroy()方法没有被调用，Activity 就不会被删除）。移到空进程目的只有一个，就是 Android 系统可以快速恢复到前台进程，而无须重新产生 Activity。

5.2.3 Activity 的生命周期（【例三】：登录响应-生命周期）

【例三】 在【例一】的基础上，程序代码中添加生命周期监测功能，再次运行程序，从开发环境界面下方的日志输出中观察程序中各个页面的生命过程。

1. 添加生命周期监测代码

（1）在登录页面 LoginActivity.java 中添加代码（加黑处），代码如下：

```
package com.easybooks.android.loginpage;
...
import android.util.Log;                                         //日志输出命名空间

public class LoginActivity extends AppCompatActivity {
    private EditText myName;
    ...
    private final static String TAG = "登录页面（LoginActivity）";    //定义显示内容变量

    @Override
```

```java
    protected void onCreate(Bundle savedInstanceState) {
        super.onCreate(savedInstanceState);
        setContentView(R.layout.login);
        Log.i(TAG, "onCreate");                          //在 onCreate()阶段显示
        findViews();
    }
    private void findViews() {    ...    }
    public void onLoginClick(View view) {    ...    }

    //Activity 生命周期监测
    @Override
    protected void onStart() {
        super.onStart();
        Log.i(TAG, "onStart");
    }

    @Override
    protected void onResume() {
        super.onResume();
        Log.i(TAG, "onResume");
    }

    @Override
    protected void onPause() {
        super.onPause();
        Log.i(TAG, "onPause");
    }

    @Override
    protected void onStop() {
        super.onStop();
        Log.i(TAG, "onStop");
    }

    @Override
    protected void onRestart() {
        super.onRestart();
        Log.i(TAG, "onRestart");
    }

    @Override
    protected void onDestroy() {
        super.onDestroy();
        Log.i(TAG, "onDestroy");
    }
}
```

（2）在欢迎页面 WelcomeActivity.java 中添加代码（加黑处），如下：
```java
package com.easybooks.android.loginpage;
...
import android.util.Log;
/**
```

```java
 * Created by Administrator on 2016/10/26.
 */
public class WelcomeActivity extends AppCompatActivity {
    private TextView myWelcome;
    private TextView myTime;
    private final static String TAG = "欢迎页面（WelcomeActivity）";

    @Override
    protected void onCreate(Bundle savedInstanceState) {
        super.onCreate(savedInstanceState);
        setContentView(R.layout.welcome);
        Log.i(TAG, "onCreate");
        findViews();
        showWelcome();
    }

    private void findViews() {    ...    }
    private void showWelcome() {    ...    }
    public void onBackClick(View view) {
        finish();
    }

    //Activity 生命周期监测
    @Override
    protected void onStart() {
        super.onStart();
        Log.i(TAG, "onStart");
    }

    @Override
    protected void onResume() {
        super.onResume();
        Log.i(TAG, "onResume");
    }

    @Override
    protected void onPause() {
        super.onPause();
        Log.i(TAG, "onPause");
    }

    @Override
    protected void onStop() {
        super.onStop();
        Log.i(TAG, "onStop");
    }

    @Override
    protected void onRestart() {
        super.onRestart();
        Log.i(TAG, "onRestart");
```

```java
    }

    @Override
    protected void onDestroy() {
        super.onDestroy();
        Log.i(TAG, "onDestroy");
    }
}
```

2. 运行效果

(1) 程序初启动，显示登录页面，观察页面生命周期如图 5.11 所示。

```
11-11 09:00:07.133 4321-4321/com.easybooks.android.loginpage I/登录页面（LoginActivity）: onStart
11-11 09:00:07.133 4321-4321/com.easybooks.android.loginpage I/登录页面（LoginActivity）: onResume
```

图 5.11　页面生命周期

(2) 用户点击"登录"按钮，跳转到欢迎页面，观察页面生命周期如图 5.12 所示。

```
11-11 09:02:10.410 4321-4321/com.easybooks.android.loginpage I/登录页面（LoginActivity）: onPause
11-11 09:02:10.533 4321-4327/com.easybooks.android.loginpage W/art: Suspending all threads took: 10.525ms
11-11 09:02:10.725 4321-4321/com.easybooks.android.loginpage I/欢迎页面（WelcomeActivity）: onCreate
11-11 09:02:10.731 4321-4321/com.easybooks.android.loginpage I/欢迎页面（WelcomeActivity）: onStart
11-11 09:02:10.731 4321-4321/com.easybooks.android.loginpage I/欢迎页面（WelcomeActivity）: onResume
11-11 09:02:10.922 4321-4335/com.easybooks.android.loginpage W/EGL_emulation: eglSurfaceAttrib not implemented
11-11 09:02:10.922 4321-4335/com.easybooks.android.loginpage W/OpenGLRenderer: Failed to set EGL_SWAP_BEHAVIOR on surface 0xa20fec40, error=EGL_SUCCESS
11-11 09:02:11.660 4321-4335/com.easybooks.android.loginpage E/Surface: getSlotFromBufferLocked: unknown buffer: 0xb4098a50
11-11 09:02:12.020 4321-4327/com.easybooks.android.loginpage W/art: Suspending all threads took: 8.035ms
11-11 09:02:12.048 4321-4321/com.easybooks.android.loginpage I/登录页面（LoginActivity）: onStop
```

图 5.12　页面生命周期

(3) 用户点击"返回"按钮，回到登录页面，观察页面生命周期如图 5.13 所示。

```
11-11 09:02:56.603 4321-4321/com.easybooks.android.loginpage I/欢迎页面（WelcomeActivity）: onPause
11-11 09:02:56.615 4321-4321/com.easybooks.android.loginpage I/登录页面（LoginActivity）: onRestart
11-11 09:02:56.615 4321-4321/com.easybooks.android.loginpage I/登录页面（LoginActivity）: onStart
11-11 09:02:56.615 4321-4321/com.easybooks.android.loginpage I/登录页面（LoginActivity）: onResume
11-11 09:02:56.686 4321-4335/com.easybooks.android.loginpage W/EGL_emulation: eglSurfaceAttrib not implement
11-11 09:02:56.686 4321-4335/com.easybooks.android.loginpage W/OpenGLRenderer: Failed to set EGL_SWAP_BEHAV
11-11 09:02:56.934 4321-4335/com.easybooks.android.loginpage E/Surface: getSlotFromBufferLocked: unknown bu
11-11 09:02:56.956 4321-4335/com.easybooks.android.loginpage D/OpenGLRenderer: endAllStagingAnimators on 0x
11-11 09:02:57.203 4321-4321/com.easybooks.android.loginpage I/欢迎页面（WelcomeActivity）: onStop
11-11 09:02:57.203 4321-4321/com.easybooks.android.loginpage I/欢迎页面（WelcomeActivity）: onDestroy
```

图 5.13　页面生命周期

5.3　Fragment（页面版块）

5.3.1　Fragment 的生命周期

创建 Fragment 非常类似于创建 Activity，只要继承 Fragment 类并改写 Fragment 生命周期的相关方法，即可让系统在特定的时候调用这些方法。需要注意的是，Fragment 组件必须依附在 Activity 上，Activity 是主，而 Fragment 是从，因此 Fragment 生命周期会直接受到所依附的 Activity 生命周期的影响。当一个 Activity 进入 onPause()状态时，所有依附在该 Activity 上的 Fragment 都会进入 onPause()状态。当该 Activity 进入 onDestroy()状态时，所有依附的 Fragment 也都会进入 onDestroy()状态。当 Activity 正在运行时，开发者可以动态地将 Fragment 加在 Activity 上，也可以删除它。

Fragment 具有与 Activity 类似的生命周期，但比 Activity 支持更多的事件回调函数。关于 Fragment 生命周期的整个流程请参见图 5.14。

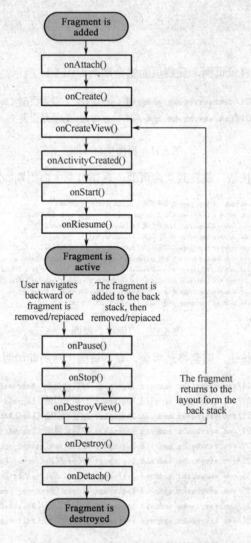

图 5.14　Fragment 生命周期的整个流程

将 Fragment 依附在 Activity 上，到画面显示会经历下列 6 个方法。

（1）onAttach()：Fragment 第一次附加在 Activity 时。
（2）onCreate()：初始化 Fragment。
（3）onCreateView()：提供 Fragment 的 UI。
（4）onActivityCreated()：Fragment 依附的 Activity 已经创建完毕。
（5）onStart()：Fragment 画面将要显示。
（6）onResume()：Fragment 画面将要与用户进行交互。

当 Fragment 画面将要离开，到脱离所依附的 Activity 会经历下列 5 个方法。

（1）onPause()：Activity 进入暂停状态或者 Fragment 将要脱离 Activity 时调用。
（2）onStop()：Activity 进入停止状态或者 Fragment 脱离 Activity 而停止时调用。
（3）onDestroyView()：Fragment 的画面确定脱离 Activity 时调用。
（4）onDestroy()：Fragment 将被卸载。
（5）onDetach()：Fragment 确定完全脱离 Activity 时调用。

通常情况下，创建 Fragment 需要继承 Fragment 的基类，并至少应实现 onCreate()、onCreateView() 和 onPause()三个生命周期的回调函数。

当然，如果仅通过 Fragment 显示元素，而不进行任何的数据保存和界面事件处理，则仅实现 onCreateView()函数也可创建 Fragment。

5.3.2　Fragment 应用（【例四】：分类预览图书）

【例四】在第 4 章【例三】预览图书页面的基础上，利用 Fragment，分成"所有"、"数据库"、"Java 开发"、"其他"页面分类显示相应的图书信息。

1. 设计页面

新建 Android 工程 Fragment。

（1）设计主页面 content_main.xml，页面上拖曳添加一个 FrameLayout 类型的控件（专用于显示版块），如图 5.15 所示。

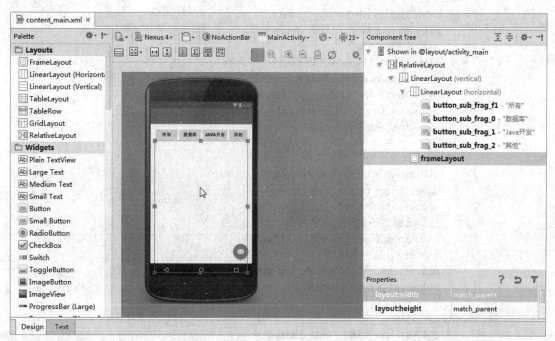

图 5.15　Fragment 设计视图

content_main.xml 源码如下：

```
<?xml version="1.0" encoding="utf-8"?>
<RelativeLayout xmlns:android="http://schemas.android.com/apk/res/android"
    xmlns:app="http://schemas.android.com/apk/res-auto"
    xmlns:tools="http://schemas.android.com/tools"
    android:layout_width="match_parent"
```

```xml
    android:layout_height="match_parent"
    android:paddingBottom="@dimen/activity_vertical_margin"
    android:paddingLeft="@dimen/activity_horizontal_margin"
    android:paddingRight="@dimen/activity_horizontal_margin"
    android:paddingTop="@dimen/activity_vertical_margin"
    app:layout_behavior="@string/appbar_scrolling_view_behavior"
    tools:context="com.easybooks.android.fragment.MainActivity"
    tools:showIn="@layout/activity_main">

    <LinearLayout
        android:orientation="vertical"
        android:layout_width="match_parent"
        android:layout_height="match_parent"
        android:layout_alignParentTop="true"
        android:layout_centerHorizontal="true">

        <LinearLayout
            android:orientation="horizontal"
            android:layout_width="wrap_content"
            android:layout_height="wrap_content">

            <Button
                android:layout_width="wrap_content"
                android:layout_height="wrap_content"
                android:text="所有"
                android:id="@+id/buttonAll"
                android:textColor="@android:color/black"
                android:textStyle="bold"
                android:textSize="@dimen/abc_text_size_menu_material"
                android:onClick="onAllClick" />

            <Button
                android:layout_width="wrap_content"
                android:layout_height="wrap_content"
                android:text="数据库"
                android:id="@+id/buttonDb"
                android:textColor="@android:color/black"
                android:textStyle="bold"
                android:textSize="@dimen/abc_text_size_menu_material"
                android:onClick="onDbClick" />

            <Button
                android:layout_width="wrap_content"
                android:layout_height="wrap_content"
                android:text="Java 开发"
                android:id="@+id/buttonJava"
                android:textColor="@android:color/black"
                android:textStyle="bold"
                android:textSize="@dimen/abc_text_size_menu_material"
                android:onClick="onJavaClick" />
```

```xml
        <Button
            android:layout_width="wrap_content"
            android:layout_height="wrap_content"
            android:text="其他"
            android:id="@+id/buttonOther"
            android:textColor="@android:color/black"
            android:textStyle="bold"
            android:textSize="@dimen/abc_text_size_menu_material"
            android:onClick="onOtherClick" />

    </LinearLayout>

    <FrameLayout
        android:layout_width="match_parent"
        android:layout_height="match_parent"
        android:id="@+id/frameLayout"></FrameLayout>

    </LinearLayout>
</RelativeLayout>
```

其中：

页面包含"所有"、"数据库"、"Java 开发"、"其他"4 个命令按钮。下部一个 Fragment，用于显示对应类型的图书信息。

（2）设计进入图书页面 enter.xml，如图 5.16 所示。

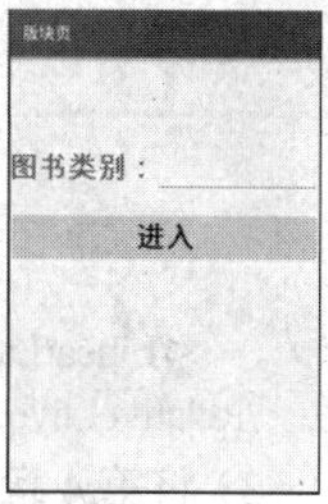

enter.xml 代码如下：

```xml
<?xml version="1.0" encoding="utf-8"?>
<RelativeLayout xmlns:android="http://schemas.android.com/apk/res/android"
    android:layout_width="match_parent"
    android:layout_height="match_parent">
    <LinearLayout
        android:orientation="vertical"
        android:layout_width="match_parent"
        android:layout_height="match_parent">
        <LinearLayout
            android:orientation="horizontal"
            android:layout_width="match_parent"
            android:layout_height="wrap_content"
            android:paddingTop="100dp">
            <TextView
                android:layout_width="wrap_content"
                android:layout_height="wrap_content"
                android:text="图书类别："
                android:id="@+id/myLabelCategory"
                android:textStyle="bold"
                android:textSize="@dimen/abc_text_size_display_1_material"
                android:layout_row="1"
                android:layout_column="0" />
            <EditText
                android:layout_width="wrap_content"
                android:layout_height="wrap_content"
                android:inputType="textPersonName"
```

图 5.16　进入图书页面

```xml
                    android:ems="10"
                    android:id="@+id/myTextCategory"
                    android:textSize="@dimen/abc_text_size_display_1_material"
                    android:layout_row="1"
                    android:layout_column="1" />
            </LinearLayout>

            <LinearLayout
                    android:orientation="horizontal"
                    android:layout_width="match_parent"
                    android:layout_height="wrap_content"
                    android:layout_columnSpan="2"
                    android:weightSum="1"
                    android:gravity="center"
                    android:paddingTop="20dp">
                <Button
                    android:layout_width="match_parent"
                    android:layout_height="wrap_content"
                    android:text="进入"
                    android:id="@+id/myButtonEnter"
                    android:textSize="@dimen/abc_text_size_display_1_material"
                    android:textStyle="bold"
                    android:layout_weight="0.5" />
            </LinearLayout>
    </LinearLayout>
</RelativeLayout>
```

（3）将第 4 章【例三】中的 book.xml 和 listitem.xml 页面文件复制到当前工程的 res\layout 目录下。

（4）将第 4 章【例三】中 res\values 目录下的字符串资源 strings.xml 和 res\drawable 目录下的图片资源复制到当前工程的对应的目录下。

2. 创建 Fragment 类

在进行 Android 版块程序设计时，每一个版块要对应创建一个 Fragment 类，其地位和作用有点儿类似于多页面程序中的 Activity 类。

（1）设计页面作为版块内容类 EnterFragment.java。

实现如下功能。

① 获得页面"图书类别"输入文本框和"进入"命令按钮的引用。

② 因为页面"进入"命令按钮没有直接定义单击事件的方法，所以需要创建监听该按钮的单击事件的方法。

③ 创建管理 Fragment 的对象（FragmentManager）实例：

FragmentManager manager = getActivity().getSupportFragmentManager();

创建 BookFragment 类实例：

BookFragment bookFragment = new BookFragment();

④ 创建 Bundle 实例，用于向图书版块传递本页面输入的显示图书分类信息。

根据在"图书类别"文本框输入的信息，变换成数值代号，以"group"标识加入接口，然后传递到图书版块中。

⑤ 创建 Fragment 的对象事务（FragmentTransaction）实例：

FragmentTransaction transaction = manager.beginTransaction();

加入"图书版块"事务：transaction.add()，参数包含需要加入的图书版块类及其标识符（"bookTag"）。然后剥离"进入版块"事务：transaction.detach()，参数包含需要剥离的进入版块的标识（"enterTag"）。而进入版块的标识是在 MainActivity()的 onCreate()中调用 loadFragment()方法加入"进入版块"事务中定义的。

最后事务确认：transaction.commit();

代码如下：

```java
package com.easybooks.android.fragment;
import …

public class EnterFragment extends Fragment {
    private EditText myCategory;
    private Button myEnter;

    @Override
    public View onCreateView(LayoutInflater inflater, ViewGroup container, Bundle savedInstanceState) {
        super.onCreateView(inflater, container, savedInstanceState);
        View view = inflater.inflate(R.layout.enter, container, false);        //显示 enter.xml 页面
        //初始化版块界面上的控件
        myCategory = (EditText) view.findViewById(R.id.myTextCategory);        //①
        myEnter = (Button) view.findViewById(R.id.myButtonEnter);              //①
        myEnter.setOnClickListener(new View.OnClickListener() {                //②
            @Override
            public void onClick(View v) {                                      //②
                String categoryname = myCategory.getText().toString();
                //在"进入版块"上启动"图书版块"
                FragmentManager manager = getActivity().getSupportFragmentManager();   //③
                BookFragment bookFragment = new BookFragment();                //③
                Bundle bundle = new Bundle();                                  //④
                switch (categoryname) {                                        //④
                    case "all":
                        bundle.putInt("group", -1);                            //④
                        break;
                    case "db":
                        bundle.putInt("group", 0);
                        break;
                    case "java":
                        bundle.putInt("group", 1);
                        break;
                    case "other":
                        bundle.putInt("group", 2);
                        break;
                    default:
                        bundle.putInt("group", -1);
                        break;
                }
                bookFragment.setArguments(bundle);                             //④
                TagString tagString = new TagString();
                FragmentTransaction transaction = manager.beginTransaction();  //③
                //加入"图书版块"
```

```
                    transaction.add(R.id.frameLayout, bookFragment, "bookTag");    //⑤
                    Fragment fragment = manager.findFragmentByTag("enterTag");     //⑤
                    //剥离"进入版块"
                    transaction.detach(fragment);                                   //⑤
                    transaction.commit();                                           //⑤
                }
            });
            return view;
        }
    }
```

（2）修改图书页面作为版块内容类 BookFragment.java。

① 将"进入"页面传递的"group"标识内容放入 curgroup 变量：

`curgroup = getArguments().getInt("group");`

② 根据 curgroup 中的图书类别代号，显示相应类别的图书信息。

代码如下：

```
package com.easybooks.android.fragment;
import ...

public class BookFragment extends Fragment{
    private ListView listView;
    private int[][] image = {
            {R.drawable.db0, R.drawable.db1, R.drawable.db2, R.drawable.db3, R.drawable.db4},
            {R.drawable.java0, R.drawable.java1, R.drawable.java2},
            {R.drawable.qt0, R.drawable.qt1, R.drawable.qt2, R.drawable.qt3, R.drawable.qt4}
    };                  //图书封面图片
    private int[][] title = {
            {R.string.title_db0, R.string.title_db1, R.string.title_db2, R.string.title_db3, R.string.title_db4},
            {R.string.title_java0, R.string.title_java1, R.string.title_java2},
            {R.string.title_qt0, R.string.title_qt1, R.string.title_qt2, R.string.title_qt3, R.string.title_qt4}
    };                  //书名
    private int[][] info = {
            {R.string.info_db0, R.string.info_db1, R.string.info_db2, R.string.info_db3, R.string.info_db4},
            {R.string.info_java0, R.string.info_java1, R.string.info_java2},
            {R.string.info_qt0, R.string.info_qt1, R.string.info_qt2, R.string.info_qt3, R.string.info_qt4}
    };                  //书号
    private int curgroup;     //(b)当前要预览的图书类别(-1:所有;0:数据库;1:Java 开发;2:其他)

    @Override
    public View onCreateView(LayoutInflater inflater, ViewGroup container, Bundle savedInstanceState) {
        super.onCreateView(inflater, container, savedInstanceState);
        View view = inflater.inflate(R.layout.book, container, false);
        listView = (ListView) view.findViewById(R.id.listView);
        findViews();
        return view;
    }

    private void findViews() {
        curgroup = getArguments().getInt("group");                  //①
        List<Map<String, Object>> list = new ArrayList<Map<String, Object>>();
        for (int i = 0; i < 3; i++) {
```

```java
            if ((curgroup != -1 && curgroup == i) || curgroup == -1) {     //②符合图书类别
                for (int j = 0; j < image[i].length; j++) {
                    Map<String, Object> item = new HashMap<String, Object>();
                    item.put("image", image[i][j]);
                    item.put("title", getResources().getString(title[i][j]));
                    item.put("info", getResources().getString(info[i][j]));
                    list.add(item);
                }
            }
        }
        SimpleAdapter simpleAdapter = new SimpleAdapter(getActivity(), list, R.layout.listitem, new String[]
{"image", "title", "info"}, new int[]{R.id.image, R.id.title, R.id.info});
        listView.setAdapter(simpleAdapter);
        //注册监听器,用消息显示用户点击的图书选项信息
        listView.setOnItemClickListener(new AdapterView.OnItemClickListener() {
            @Override
            public void onItemClick(AdapterView<?> parent, View view, int position, long id) {
                String name = "", isbn = "";
                int row;
                if (curgroup != -1) {                    //②
                    row = curgroup;                      //②
                    name = getString(title[row][position]);
                    isbn = getString(info[row][position]);
                } else {
                    if (position < 5) {
                        row = 0;
                        name = getString(title[row][position]);
                        isbn = getString(info[row][position]);
                    } else if (position < 8) {
                        row = 1;
                        name = getString(title[row][position - 5]);
                        isbn = getString(info[row][position - 5]);
                    } else if (position < 13) {
                        row = 2;
                        name = getString(title[row][position - 8]);
                        isbn = getString(info[row][position - 8]);
                    }
                }
                Toast.makeText(getActivity(), "您选购了:《" + name + "》, " + isbn, Toast.LENGTH_SHORT).show();
            }
        });
    }
}
```

3. 主程序功能

在 MainActivity.java 中编写代码,实现初始加载和更换不同的版块功能。

(1) 初始加载"进入版块",通过在 onCreate 中调用 loadFragment()方法实现。
(2) 更换不同的图书类别版块

```java
public void previewBook(int group){     …     }
```

其中不同的命令按钮，代入不同的图书类别参数给 group，将其值传递给图书分类版块显示。
MainActivity.java 代码如下：

```java
package com.easybooks.android.fragment;
import …
public class MainActivity extends AppCompatActivity {

    @Override
    protected void onCreate(Bundle savedInstanceState) {
        super.onCreate(savedInstanceState);
        setContentView(R.layout.activity_main);
        loadFragment();                                  //(1)
        Toolbar toolbar = (Toolbar) findViewById(R.id.toolbar);
        setSupportActionBar(toolbar);

        FloatingActionButton fab = (FloatingActionButton) findViewById(R.id.fab);
        fab.setOnClickListener(new View.OnClickListener() {
            @Override
            public void onClick(View view) {
                Snackbar.make(view, "Replace with your own action", Snackbar.LENGTH_LONG)
                        .setAction("Action", null).show();
            }
        });
    }

    public void loadFragment() {                         //(1)
        //初始时加载"进入版块"
        FragmentManager manager = getSupportFragmentManager();
        EnterFragment enterFragment = new EnterFragment();
        FragmentTransaction transaction = manager.beginTransaction();
        transaction.add(R.id.frameLayout, enterFragment, "enterTag");
        transaction.commit();
    }

    public void onButtonAllClick(View view) {            //单击"所有"命令按钮
        previewBook(-1);                                 //显示-1 类型图书
    }

    public void onButtonDbClick(View view) {             //单击"数据库"命令按钮
        previewBook(0);                                  //显示 0 类型图书
    }

    public void onButtonJavaClick(View view) {           //单击"Java 开发"命令按钮
        previewBook(1);                                  //显示 1 类型图书
    }

    public void onButtonOtherClick(View view) {          //单击"其他"命令按钮
        previewBook(2);                                  //显示 2 类型图书
    }

    public void previewBook(int group) {                 //(2)
```

```java
//根据用户点击按钮选择的类别,向"图书版块"传递 group 参数以显示不同类别书的预览
FragmentManager manager = getSupportFragmentManager();
FragmentTransaction transaction = manager.beginTransaction();
BookFragment bookFragment = new BookFragment();
Bundle bundle = new Bundle();
bundle.putInt("group", group);
bookFragment.setArguments(bundle);
//用对应类别的"图书版块"取代原版块
transaction.replace(R.id.frameLayout, bookFragment, "bookTag");
transaction.commit();
}

@Override
public boolean onCreateOptionsMenu(Menu menu) {
    // Inflate the menu; this adds items to the action bar if it is present.
    getMenuInflater().inflate(R.menu.menu_main, menu);
    return true;
}

@Override
public boolean onOptionsItemSelected(MenuItem item) {
    // Handle action bar item clicks here. The action bar will
    // automatically handle clicks on the Home/Up button, so long
    // as you specify a parent activity in AndroidManifest.xml.
    int id = item.getItemId();

    //noinspection SimplifiableIfStatement
    if (id == R.id.action_settings) {
        return true;
    }

    return super.onOptionsItemSelected(item);
}
}
```

4. 运行效果

运行效果如图 5.17 至图 5.19 所示。

图 5.17 在初始页面上输入"db"

图 5.18 单击"进入"后页面

Android 实用教程（基于 Android Studio·含视频分析）

图 5.19 单击"JAVA 开发"后选择"JavaEE 基础实用教程（第 2 版）"

5.3.3 Fragment 生命周期（【例五】：分类预览图书-生命周期）

【例五】在【例四】"分类预览图书"的基础上，程序代码中添加版块生命周期监测功能，再次运行程序，从开发环境界面下方的日志输出中观察程序中各个版块的生命过程。

1. 添加生命周期监测代码

（1）创建 TagString 类用于统一管理各个 Fragment 类的标识。

TagString.java 代码如下：

```
package com.easybooks.android.fragment;
public class TagString {
    private final static String TAG_Ent = "进入版块（EnterFragment）";
    private final static String TAG_Book = "图书版块（BookFragment）";

    public String getTAG_Ent() {
        return TAG_Ent;
    }

    public String getTAG_Book() {
        return TAG_Book;
    }
}
```

（2）在 MainActivity.java 类 onCreate 中调用的 loadFragment()方法中加入代码，获得进入版块的标识，在加入 Fragment 事务时将进入版块（enterFragment）与进入版块的标识关联。

```
public void loadFragment() {
    //初始时加载"进入版块"
```

```
        FragmentManager manager = getSupportFragmentManager();
        FragmentTransaction transaction = manager.beginTransaction();
        EnterFragment enterFragment = new EnterFragment();
        TagString tagString = new TagString();
        transaction.add(R.id.frameLayout, enterFragment, tagString.getTAG_Ent());
        transaction.commit();
}
```

（3）在进入（EnterFragment.java）和图书版块（BookFragment.java）对应的 Fragment 类中添加监测版块生命周期的代码（添加的代码完全一样），如下：

```
public class EnterFragment extends Fragment {
    …
    private static String TAG;
    …


    @Override
public View onCreateView(LayoutInflater inflater, ViewGroup container, Bundle savedInstanceState) {
    super.onCreateView(inflater, container, savedInstanceState);
    TagString tag = new TagString();
    TAG = tag.getTAG_Ent();      //图书版块 TAG = tag.getTAG_Book();
    Log.i(TAG, "onCreateView");
    View view = inflater.inflate(R.layout.login, container, false);
    //初始化版块界面上的各控件
        ...
    return view;
    }

    //Fragment 生命周期监测
    @Override
    public void onStart() {
        super.onStart();
        Log.i(TAG, "onStart");
    }

    @Override
    public void onResume() {
        super.onResume();
        Log.i(TAG, "onResume");
    }

    @Override
    public void onPause() {
        super.onPause();
        Log.i(TAG, "onPause");
    }

    @Override
    public void onStop() {
        super.onStop();
        Log.i(TAG, "onStop");
    }
```

```java
    @Override
    public void onDestroy() {
        super.onDestroy();
        Log.i(TAG, "onDestroy");
    }

    @Override
    public void onAttach(Activity activity){
        super.onAttach(getActivity());
        Log.i(TAG,"onAttach");
    }

    @Override
    public void onDetach(){
        super.onDetach();
        Log.i(TAG,"onDetach");
    }
}
```

2. 运行效果

运行效果：如图 5.20 至图 5.22 所示。

```
12-27 09:10:09.351 22345-22345/com.easybooks.android.fragment I/进入版块（EnterFragment）: onCreateView
12-27 09:10:09.353 22345-22345/com.easybooks.android.fragment I/进入版块（EnterFragment）: onStart
12-27 09:10:09.354 22345-22345/com.easybooks.android.fragment I/进入版块（EnterFragment）: onResume
```

图 5.20　程序初启动，显示进入版块

```
12-27 09:25:46.935 836-836/com.easybooks.android.fragment I/进入版块（EnterFragment）: onPause
12-27 09:25:46.935 836-836/com.easybooks.android.fragment I/进入版块（EnterFragment）: onStop
12-27 09:25:46.936 836-836/com.easybooks.android.fragment I/图书版块（BookFragment）: onCreateView
12-27 09:25:46.938 836-836/com.easybooks.android.fragment I/图书版块（BookFragment）: onStart
12-27 09:25:46.938 836-836/com.easybooks.android.fragment I/图书版块（BookFragment）: onResume
```

图 5.21　单击"进入"按钮，进入图书版块

```
12-27 09:27:24.110 836-836/com.easybooks.android.fragment I/图书版块（BookFragment）: onAttach
12-27 09:27:24.110 836-836/com.easybooks.android.fragment I/图书版块（BookFragment）: onPause
12-27 09:27:24.110 836-836/com.easybooks.android.fragment I/图书版块（BookFragment）: onStop
12-27 09:27:24.110 836-836/com.easybooks.android.fragment I/图书版块（BookFragment）: onDestroy
12-27 09:27:24.110 836-836/com.easybooks.android.fragment I/图书版块（BookFragment）: onDetach
12-27 09:27:24.110 836-836/com.easybooks.android.fragment I/图书版块（BookFragment）: onCreateView
12-27 09:27:24.111 836-836/com.easybooks.android.fragment I/图书版块（BookFragment）: onStart
12-27 09:27:24.111 836-836/com.easybooks.android.fragment I/图书版块（BookFragment）: onResume
```

图 5.22　用户点击界面上方各标签按钮进入对应类别的图书版块

第 6 章 Android 用户界面进阶

6.1 菜 单

菜单是应用程序中非常重要的组成部分，为程序开发人员提供易于使用的编程接口。能够在不占用界面空间的前提下，提供统一的功能选择界面。Android 系统支持三种菜单模式，分别是选项菜单（Option Menu）、子菜单（Sub Menu）和快捷菜单（Context Menu）。

Android 菜单有以下两种开发方式。

（1）使用 XML 文件制作菜单资源，然后通过 inflate()函数映射到程序代码中。

（2）直接在代码中定义菜单，由程序在运行时动态生成。

注意，因为本章菜单选项所指向的功能页包括登录、注册、图书等前面已经设计的页面，读者需要将它们复制到当前所做工程对应的目录下，Java 源文件还需要把命名空间修改为当前工程的命名空间。

6.1.1 选项菜单（【例一】：调用第 4 章例二、例三和第 5 章例一）

选项菜单是垂直的列表型浮动菜单，如图 6.1 所示。浮动菜单不能够显示图标，但支持单选框和复选框，原来图标菜单的部分功能由操作栏代替实现。

【例一】用 XML 文件描述菜单和用代码定义生成选项菜单，分别调用第 4 章例二注册、第 5 章例一登录响应和第 4 章例三图书展示功能。

1. XML 文件描述菜单

用 XML 文件描述菜单是较好的选择，可以将菜单的内容与功能代码分离，且有利于分析和调整菜单结构。

（1）准备资源

创建 Android 工程 OptionMenu，准备以下资源。

第 4 章【例二】：register.xml、RegisterActivity.java；

第 4 章【例三】：book.xml、listitem.xml、BookActivity.java。

第 5 章【例一】：login.xml、welcome.xml、LoginActivity.java、WelcomeActivity.java。

图 6.1　浮动菜单

其中，所有 xml 文件都复制到项目\app\src\main\res\layout 目录下，所有.java 文件则复制到项目\app\src\main\java\com\easybooks\android\optionmenu 目录下，并把命名空间改为"package com.easybooks.android.optionmenu;"，同时注意检查 import。

把第 4 章例三和第 5 章例一的图片资源复制到项目\app\src\main\res\drawable 目录下。

把 strings.xml 换成第 4 章例三的内容。

（2）描述菜单

Android 程序的菜单资源默认存放在项目的\app\src\main\res\menu 目录下，在该路径下面有一个名为"menu_main.xml"的文件（见图 6.2），用户可以在其中编写代码来描述自己所设计的菜单。

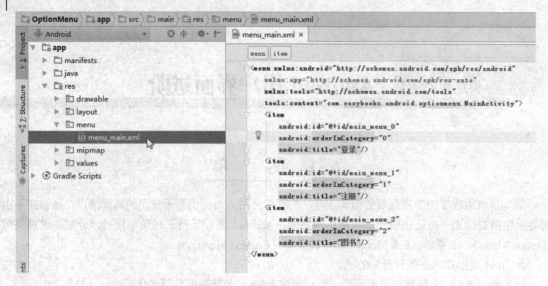

图 6.2 菜单资源文件

本例中 menu_main.xml 文件生成具有 3 个子项的菜单，代码如下：

```xml
<menu xmlns:android=http://schemas.android.com/apk/res/android
    xmlns:app="http://schemas.android.com/apk/res-auto"
    xmlns:tools="http://schemas.android.com/tools"
    tools:context="com.easybooks.android.optionmenu.MainActivity">
    <item
        android:id="@+id/main_menu_0"
        android:orderInCategory="0"
        android:title="登录"/>
    <item
        android:id="@+id/main_menu_1"
        android:orderInCategory="1"
        android:title="注册"/>
    <item
        android:id="@+id/main_menu_2"
        android:orderInCategory="2"
        android:title="图书"/>
</menu>
```

其中，menu 是菜单的容器，菜单资源必须以 menu 作为根元素。item 是菜单项，其属性值 id 和 title 分别是菜单项的 ID 值和标题。

（3）代码映射

在 MainActivity.java 中修改代码，创建 menu_main.xml 菜单文件后，系统在 MainActivity 中生成菜单资源初始化代码，用户需要取代系统生成的代码如下。

```java
package com.easybooks.android.optionmenu;
import …
public class MainActivity extends AppCompatActivity {

    @Override
    protected void onCreate(Bundle savedInstanceState) {    ...    }

    @Override
```

```java
        public boolean onCreateOptionsMenu(Menu menu) {                    //①
            getMenuInflater().inflate(R.menu.menu_main, menu);
            return true;
        }

        @Override
        public boolean onOptionsItemSelected(MenuItem item) {              //②
            Intent intent;
            switch (item.getItemId()) {                                    //③
                case R.id.main_menu_0:
                    intent = new Intent(this, LoginActivity.class);
                    startActivity(intent);
                    return true;
                case R.id.main_menu_1:
                    intent = new Intent(this, RegisterActivity.class);
                    startActivity(intent);
                    return true;
                case R.id.main_menu_2:
                    intent = new Intent(this, BookActivity.class);
                    startActivity(intent);
                    return true;
                default:
                    return false;
            }
        }
    }
```

说明：

① 在 Android 4.0 系统中，Activity 在创建时便会调用 onCreateOptionsMenu()函数初始化自身的菜单系统。在 Activity 的整个生命周期中，选项菜单是一直被重复利用的，直到 Activity 被销毁。重载 onCreateOptionsMenu()函数主要目的是初始化菜单，可以使用 XML 文件的菜单资源，也可以使用代码动态加载菜单。

② 在用户选择菜单项后，Android 系统会调用 onOptionsItemSelected()函数，一般将菜单选择事件的响应代码放置在该函数中，它返回用户选择的 MenuItem。

可以通过 getItemId()函数获取 MenuItem 的 ID，这个 ID 就是用户在 XML 文件中为每个菜单项所设定的 android:id 属性值。函数 onOptionsItemSelected()的返回值表示是否需要其他事件处理函数对菜单选择事件进行处理，如果不需要其他函数处理该事件则返回 true，否则返回 false。

③ getItemId()函数获取被选择菜单子项的 ID，通过 case 项进行匹配，通过接口连接到相应的类处理。

（4）配置 Activity

在 AndroidManifest.xml 文件中，配置（加黑处）如下：

```xml
<?xml version="1.0" encoding="utf-8"?>
<manifest xmlns:android="http://schemas.android.com/apk/res/android"
    package="com.easybooks.android.optionmenu">

    <application
        ...
        <activity
```

```
                android:name=".MainActivity"
            ...
        </activity>
        <activity
            android:name=".LoginActivity"
            android:label="登录"/>
        <activity
            android:name=".WelcomeActivity"
            android:label="欢迎"/>
        <activity
            android:name=".RegisterActivity"
            android:label="注册"/>
        <activity
            android:name=".BookActivity"
            android:label="新书"/>
    </application>
</manifest>
```

（5）运行菜单

程序运行后，系统显示如图6.3所示。

图6.3 初始页面

实际设计时用户可以修改该页面内容为用户初始页面。点击后面三个点，系统调出选项菜单。点选不同的菜单项可以分别启动登录、注册、新书等不同的页面，如图6.4所示。

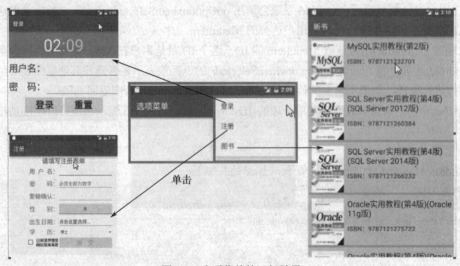

图6.4 选项菜单的运行效果

进入登录页面后，输入有关的信息，单击"登录"，系统显示"欢迎"页面。

2. 用代码定义生成菜单

（1）创建 Android 工程 OptionMenubyCode。
（2）准备资源与 OptionMenu 完全相同，但不用编写 menu_main.xml 文件。
（3）在 MainActivity.java 中编写代码来实现菜单。

首先要在代码中定义菜单 ID，然后在 onCreateOptionsMenu()函数中添加选项菜单，并设置菜单的标题等信息。

在 MainActivity.java 修改和编写代码如下：

```
package com.easybooks.android.optionmenubycode;
import android.content.Intent;
...
public class MainActivity extends AppCompatActivity {
    final static int MENU_00 = Menu.FIRST;                //①
    final static int MENU_01 = Menu.FIRST + 1;            //①
    final static int MENU_02 = Menu.FIRST + 2;            //①

    @Override
    protected void onCreate(Bundle savedInstanceState) {  ...  }

    @Override
    public boolean onCreateOptionsMenu(Menu menu) {       //②
        menu.add(0, MENU_00, 0, "登录");                   //②
        menu.add(0, MENU_01, 1, "注册");
        menu.add(0, MENU_02, 2, "图书");

        return true;
    }

    @Override
    public boolean onOptionsItemSelected(MenuItem item) {
        Intent intent;
        switch (item.getItemId()) {
            case MENU_00:
                intent = new Intent(this, LoginActivity.class);
                startActivity(intent);
                return true;
            case MENU_01:
                intent = new Intent(this, RegisterActivity.class);
                startActivity(intent);
                return true;
            case MENU_02:
                intent = new Intent(this, BookActivity.class);
                startActivity(intent);
                return true;
            default:
                return false;
        }
    }
}
```

说明：

① 一般将菜单项的 ID 定义成静态常量，并使用静态常量 Menu.FIRST（整数类型，值为 1）定义第一个菜单子项，以后的菜单项在 Menu.FIRST 增加相应的数值。

② 在 onCreateOptionsMenu()函数中，函数的返回类型为布尔值，返回 true 则可显示在函数中设置的菜单，否则将不能够显示菜单。Menu 对象作为一个参数被传递到函数内部，因此在 onCreateOptionsMenu()函数中，用户可以使用 Menu 对象的 add()函数添加菜单项。

```
Menu.add (int groupId,   int itemId,   int order,   CharSequencetitle)
```

其中：

groupId：组 ID，用以批量地对菜单子项进行处理和排序。

ItemId：子项 ID，是每一个菜单子项的唯一标识。通过子项 ID 使应用程序能够定位到用户所选择的菜单子项。

order：定义菜单子项在选项菜单中的排列顺序。

title：菜单子项所显示的标题。

（4）配置 Activity

在 AndroidManifest.xml 文件中配置 Activity，内容与项目 OptionMenu 的完全一样。

（5）运行菜单

程序运行后，系统显示如图 6.4 所示。

6.1.2 子菜单（【例二】：第 4 章例二、例三组和第 5 章例一分类组）

当程序具有大量的功能时，可以将相似的功能划分成组，选项菜单可用来表示功能组，而具体功能则可由子菜单进行选择。子菜单就是二级菜单，用户点击选项菜单或快捷菜单中的菜单项，就可以打开子菜单。

传统的子菜单一般采用树形的层次化结构，但 Android 系统却使用浮动窗体的形式显示菜单子项。采用与众不同的显示方式，主要是为了更好地适应小屏幕的显示方式。子菜单不支持嵌套，也就是说不能够在子菜单中再使用子菜单。

【例二】将第 4 章例二注册和第 5 章例一登录功能作为"用户"组，再将第 4 章例三图书展示变成分类展示，下面分别说明 XML 文件描述子菜单和用代码定义生成子菜单。

1. XML 文件描述

（1）创建 Android 子菜单工程 SubMenu。准备子菜单资源：需要资源与 OptionMenubyCode 完全相同，包括 strings.xml 文件。

另外加入用作子菜单项的图标文件：allicon.png、dbicon.png、javaicon.png、qticon.png、log.png 和 reg.png 放入 res\drawable 目录。

（2）描述菜单

系统生成的 menu_main.xml 文件的代码如下：

```xml
<menu xmlns:android=http://schemas.android.com/apk/res/android
    xmlns:app="http://schemas.android.com/apk/res-auto"
    xmlns:tools="http://schemas.android.com/tools"
    tools:context="com.easybooks.android.submenu.MainActivity">
    <item
        android:id="@+id/action_settings"
        android:orderInCategory="100"
        android:title="@string/action_settings"
```

```
        app:showAsAction="never" />
</menu>
```

删除<item ... />内容，加入下列代码：

```
<item                                    //第1个菜单项：用户
    android:id="@+id/main_menu_0"
    android:orderInCategory="0"
    android:title="用户">
    <menu>
        <item                            //用户：第1个子菜单：登录
            android:id="@+id/sub_menu_0_0"
            android:icon="@drawable/log"
            android:title="登录"/>
        <item                            //用户：第2个子菜单：注册
            android:id="@+id/sub_menu_0_1"
            android:icon="@drawable/reg"
            android:title="注册"/>
    </menu>
</item>
<item                                    //第2个菜单项：图书
    android:id="@+id/main_menu_1"
    android:orderInCategory="1"
    android:title="图书">
    <menu>
        <item                            //图书：第1个子菜单：所有
            android:id="@+id/sub_menu_1_0"
            android:icon="@drawable/allicon"
            android:title="所有"/>
        <item                            //图书：第1个子菜单：数据库
            android:id="@+id/sub_menu_1_1"
            android:icon="@drawable/dbicon"
            android:title="数据库"/>
        <item                            //图书：第1个子菜单：Java 开发
            android:id="@+id/sub_menu_1_2"
            android:icon="@drawable/javaicon"
            android:title="Java 开发"/>
        <item                            //图书：第1个子菜单：其他
            android:id="@+id/sub_menu_1_3"
            android:icon="@drawable/qticon"
            android:title="其他"/>
    </menu>
</item>
```

子菜单也使用<menu>标签进行声明，内部使用<item>标签描述菜单项。

（3）代码映射

在 MainActivity.java 中，修改、编写代码如下：

```
package com.easybooks.android.submenu;
import android.content.Intent;
...
public class MainActivity extends AppCompatActivity {
    @Override
    protected void onCreate(Bundle savedInstanceState) {   ...    }
```

```java
@Override
public boolean onCreateOptionsMenu(Menu menu) {
    getMenuInflater().inflate(R.menu.menu_main, menu);
    return true;
}

@Override
public boolean onOptionsItemSelected(MenuItem item) {
    Intent intent;
    Bundle bundle = new Bundle();
    switch (item.getItemId()) {
        case R.id.sub_menu_0_0:
            intent = new Intent(this, LoginActivity.class);
            startActivity(intent);
            return true;
        case R.id.sub_menu_0_1:
            intent = new Intent(this, RegisterActivity.class);
            startActivity(intent);
            return true;
        case R.id.sub_menu_1_0:
            intent = new Intent(this, BookActivity.class);
            bundle.putInt("group", -1);
            intent.putExtras(bundle);
            startActivity(intent);
            return true;
        case R.id.sub_menu_1_1:
            intent = new Intent(this, BookActivity.class);
            bundle.putInt("group", 0);
            intent.putExtras(bundle);
            startActivity(intent);
            return true;
        case R.id.sub_menu_1_2:
            intent = new Intent(this, BookActivity.class);
            bundle.putInt("group", 1);
            intent.putExtras(bundle);
            startActivity(intent);
            return true;
        case R.id.sub_menu_1_3:
            intent = new Intent(this, BookActivity.class);
            bundle.putInt("group", 2);
            intent.putExtras(bundle);
            startActivity(intent);
            return true;
        default:
            return false;
    }
}
```

（4）修改 BookActivity.java

因为原来 BookActivity.java 显示所有图书列表，所以需要修改 BookActivity.java 文件，根据菜单传递的值对应的图书类型，仅仅显示对应该类型的图书列表。

修改后 BookActivity.java 代码如下：

```java
package com.easybooks.android.submenu;

import ...
public class BookActivity extends AppCompatActivity {
    private ListView myListView;

    private int[][] image = {
            {R.drawable.db0, R.drawable.db1, R.drawable.db2, R.drawable.db3, R.drawable.db4},
            {R.drawable.java0, R.drawable.java1, R.drawable.java2},
            {R.drawable.qt0, R.drawable.qt1, R.drawable.qt2, R.drawable.qt3, R.drawable.qt4}
    };                            //图书封面图片数组
    private int[][] title = {
            {R.string.title_db0, R.string.title_db1, R.string.title_db2, R.string.title_db3, R.string.title_db4},
            {R.string.title_java0, R.string.title_java1, R.string.title_java2},
            {R.string.title_qt0, R.string.title_qt1, R.string.title_qt2, R.string.title_qt3, R.string.title_qt4}
    };                            //书名数组
    private int[][] info = {
            {R.string.info_db0, R.string.info_db1, R.string.info_db2, R.string.info_db3, R.string.info_db4},
            {R.string.info_java0, R.string.info_java1, R.string.info_java2},
            {R.string.info_qt0, R.string.info_qt1, R.string.info_qt2, R.string.info_qt3, R.string.info_qt4}
    };                            //书号数组
    private int curgroup;         //当前要预览的图书类别(-1:所有;0:数据库;1:Java 开发;2:其他)
    @Override
    protected void onCreate(Bundle savedInstanceState) {
        super.onCreate(savedInstanceState);
        setContentView(R.layout.book);
        findViews();
    }

    private void findViews() {
        Bundle bundle = getIntent().getExtras();
        curgroup = bundle.getInt("group");
        myListView = (ListView) findViewById(R.id.listView);
        List<Map<String, Object>> list = new ArrayList<Map<String, Object>>();
        for (int i = 0; i < 3; i++) {
            if ((curgroup != -1 && curgroup == i) || curgroup == -1) {
                for (int j = 0; j < image[i].length; j++) {
                    Map<String, Object> item = new HashMap<String, Object>();
                    item.put("image", image[i][j]);
                    item.put("title", getResources().getString(title[i][j]));
                    item.put("info", getResources().getString(info[i][j]));
                    list.add(item);
                }
            }
        }
        SimpleAdapter simpleAdapter = new SimpleAdapter(this, list, R.layout.listitem, new String[]{"image", "title", "info"}, new int[]{R.id.image, R.id.title, R.id.info});
        myListView.setAdapter(simpleAdapter);
    }
}
```

定义图书封面图片(image)、书名(title)、书号(info) 数组

（5）配置 Activity

在 AndroidManifest.xml 文件中配置 Activity，内容与项目 OptionMenu 的完全一样。

（6）运行菜单

程序运行，如图 6.5 所示。

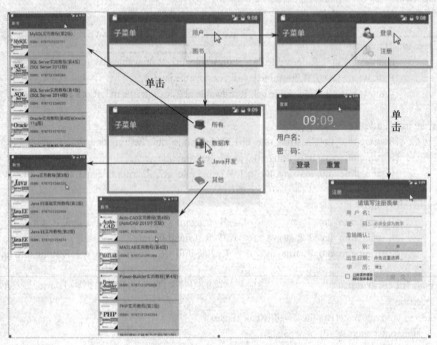

图 6.5 子菜单运行效果

Android 系统的子菜单使用起来非常灵活，除了可以用 XML 文件描述菜单结构，也可以通过代码在选项菜单或快捷菜单中使用子菜单。

2. 用代码定义生成

创建 Android 工程 SubMenubyCode。准备资源与 SubMenu 完全相同，直接在 MainActivity.java 中编写代码来实现子菜单。

在 MainActivity.java 中，修改、编写代码如下：

```java
package com.easybooks.android.submenubycode;
import …

public class MainActivity extends AppCompatActivity {
    final static int MENU_00 = Menu.FIRST;                    //①定义选项菜单 ID
    final static int MENU_01 = Menu.FIRST + 1;
    final static int SUB_MENU_00_00 = Menu.FIRST + 2;         //②定义子菜单项的 ID
    final static int SUB_MENU_00_01 = Menu.FIRST + 3;
    final static int SUB_MENU_01_00 = Menu.FIRST + 4;
    final static int SUB_MENU_01_01 = Menu.FIRST + 5;
    final static int SUB_MENU_01_02 = Menu.FIRST + 6;
    final static int SUB_MENU_01_03 = Menu.FIRST + 7;

    @Override
```

```java
    protected void onCreate(Bundle savedInstanceState) {  ...  }

    @Override
    public boolean onCreateOptionsMenu(Menu menu) {
        SubMenu subMenu1 = (SubMenu) menu.addSubMenu(0, MENU_00, 0, "用户");
            subMenu1.add(0, SUB_MENU_00_00, 0, "登录").setIcon(R.drawable.log);            //③
            subMenu1.add(0, SUB_MENU_00_01, 0, "注册").setIcon(R.drawable.reg);            //③
        SubMenu subMenu2 = (SubMenu) menu.addSubMenu(0, MENU_01, 1, "图书");
            subMenu2.add(0, SUB_MENU_01_00, 0, "所有").setIcon(R.drawable.allicon);        //③
            subMenu2.add(0, SUB_MENU_01_01, 0, "数据库").setIcon(R.drawable.dbicon);       //③
            subMenu2.add(0, SUB_MENU_01_02, 0, "Java 开发").setIcon(R.drawable.javaicon);  //③
            subMenu2.add(0, SUB_MENU_01_03, 0, "其他").setIcon(R.drawable.qticon);         //③
        return true;
    }

    @Override
    public boolean onOptionsItemSelected(MenuItem item) {
        Intent intent;
        Bundle bundle = new Bundle();
        switch (item.getItemId()) {
            case SUB_MENU_00_00:
                intent = new Intent(this, LoginActivity.class);
                startActivity(intent);
                return true;
            case SUB_MENU_00_01:
                intent = new Intent(this, RegisterActivity.class);
                startActivity(intent);
                return true;
            case SUB_MENU_01_00:
                intent = new Intent(this, BookActivity.class);
                bundle.putInt("group", -1);
                intent.putExtras(bundle);
                startActivity(intent);
                return true;
            case SUB_MENU_01_01:
                intent = new Intent(this, BookActivity.class);
                bundle.putInt("group", 0);
                intent.putExtras(bundle);
                startActivity(intent);
                return true;
            case SUB_MENU_01_02:
                intent = new Intent(this, BookActivity.class);
                bundle.putInt("group", 1);
                intent.putExtras(bundle);
                startActivity(intent);
                return true;
            case SUB_MENU_01_03:
                intent = new Intent(this, BookActivity.class);
                bundle.putInt("group", 2);
                intent.putExtras(bundle);
                startActivity(intent);
```

```
                    return true;
            default:
                    return false;
        }
    }
}
```

其中：

① 定义选项菜单和子菜单所有菜单项的 ID。

② 使用 addSubMenu()函数在选项菜单中增加了 1 个菜单项 MENU_00，当用户点击这个菜单项后会打开子菜单。共有 4 个参数。

第 1 个参数为组 ID：如果不分组则可以使用 0。

第 2 个参数为菜单项的 ID。

第 3 个参数为显示排序：数字越小越靠近列表上方。

第 4 个参数为菜单项显示的标题。

③ 在子菜单中添加了菜单项，设置了菜单的图标。

3．程序运行

程序运行结果与用 XML 文件描述的子菜单完全相同，见图 6.5。

6.1.3 快捷菜单（【例三】：根据第 4 章例三选择图书显示详细信息）

快捷菜单类似于计算机程序中的"右键菜单"，当用户点击界面上某个元素超过 2 秒后，将启动注册到该界面元素的快捷菜单。快捷菜单同样采用浮动的显示方式，虽然快捷菜单的显示方式与子菜单相同，但两种菜单的启动方式却截然不同。

【例三】在第 4 章例三图书展示的基础上，加入页面显示某本书的详细信息功能，通过图书展示页面快捷菜单"立即购买、加入购物车、详细信息"等 3 个菜单，选择"立即购买"和"加入购物车"菜单项，用显示对应图书名称、ISBN 和价格消息框模拟，选择"详细信息"，显示对应该图书详细信息页面。

实现思路如下。

（1）修改 BookActivity.java 类，实现下列功能。

将监听列表单击按钮事件变成长时间监听列表单击按钮事件：

```
setOnItemLongClickListener(new AdapterView.OnItemLongClickListener() {
    @Override
    public boolean onItemLongClick(AdapterView<?> parent, View view, int position, long id) {
        …
    })
}
```

将该图书列表的顺序号变成分类图书数组的 2 维下标。

（2）长时间单击图书列表，产生下列事件。

```
myListView.setOnCreateContextMenuListener(new View.OnCreateContextMenuListener() {
    @Override
    public void onCreateContextMenu(ContextMenu menu, View v, ContextMenu.ContextMenuInfo menuInfo) {
        menu.setHeaderTitle( … );
        menu.add( … );
        …
```

        })
    }

其中，onCreateContextMenu()函数初始化菜单项，包括添加快捷菜单所显示的标题、图标和菜单子项等内容。

第 1 个参数 menu 是需要显示的快捷菜单。

第 2 个参数 v 是用户点击的界面元素。

第 3 个参数 menuInfo 是所选择界面元素的额外信息。

setHeaderTitle()函数设置快捷菜单标题。

add()函数和 addSubMenu()函数，可以在快捷菜单中添加菜单子项和子菜单。这里，加入"立即购买"、"加入购物车"和"详细信息"等 3 项菜单项。

（3）用户选择快捷菜单的菜单项后 onContextItemSelected()函数响应菜单选择事件。

判断选择快捷菜单的菜单项：

如果是立即购买，显示图书名。

如果是加入购物车，显示图书 ISBN。

如果是详细信息，将分类图书数组的二维下标向图书详细信息页面传递，然后开启 DetailsActivity.java 类对应的页面。

（4）新建图书详细信息页面，接受图书列表页面传递的图书数组的二维下标信息，读取图书数组信息，在界面上显示出来。

1. 准备资源

将第 4 章【例三】工程（ch4\BookPage 目录）复制到 ch6 文件夹下，打开工程。

在 strings.xml 文件中为每个书的信息条目增加两个字段（name 属性名形如 price_xxx 和 details_xxx），分别录入该书的价格、内容简介的详细信息。

例如：

```
<string name="title_db0">MySQL 实用教程(第 2 版)</string>
<string name="info_db0">ISBN：9787121232701</string>
<string name="price_db0">¥53.00</string>
<string name="details_db0">以当前最流行 MySQL5.6 作为平台，分为 4 个部分，它们分别是 MySQL 综述（含习题）、MySQL 实验、综合应用练习和附录。在系统介绍 MySQL 功能的基础上，通过实验进行操作练习和消化理解。同时系统介绍目前最流行的 PHP、ASP.NET(C#)、JavaEE、Visual C++和 Visual Basic 等应用 MySQL 数据库的方法。通过本书学习模仿，基本掌握了当前几个流行平台开发 MySQL 数据库应用系统的方法，比较好地解决了 MySQL "学"和"用"的问题。本书可作为大学本科、高职高专有关课程教材，也可供广大数据库应用开发人员使用或参考。</string>
```

2. 修改图书列表页并加入快捷菜单

在 BookActivity.java 文件中定义和实现快捷菜单的功能。

对 BookActivity.java 文件进行修改、编辑的代码如下：

```java
package com.easybooks.android.bookpage;
import ...

public class BookActivity extends AppCompatActivity {
    private ListView myListView;
    private int row = 0;
    private int col = 0;

    final static int CONTEXT_MENU_00 = Menu.FIRST;
```

```java
        final static int CONTEXT_MENU_01 = Menu.FIRST + 1;
        final static int CONTEXT_MENU_02 = Menu.FIRST + 2;
```

定义图书封面图片(image)、书名(title)、书号(info)数组

```java
    @Override
    protected void onCreate(Bundle savedInstanceState) {
        super.onCreate(savedInstanceState);
        setContentView(R.layout.book);
        findViews();
    }

    private void findViews() {
        myListView = (ListView) findViewById(R.id.listView);
        List<Map<String, Object>> list = new ArrayList<Map<String, Object>>();
        for (int i = 0; i < 3; i++) {
            for (int j = 0; j < image[i].length; j++) {
                Map<String, Object> item = new HashMap<String, Object>();
                item.put("image", image[i][j]);
                item.put("title", getResources().getString(title[i][j]));
                item.put("info", getResources().getString(info[i][j]));
                list.add(item);
            }
        }
        SimpleAdapter simpleAdapter = new SimpleAdapter(this, list, R.layout.listitem, new String[]{"image", "title", "info"}, new int[]{R.id.image, R.id.title, R.id.info});
        myListView.setAdapter(simpleAdapter);
        //长按获取书目数组二维下标
        myListView.setOnItemLongClickListener(new AdapterView.OnItemLongClickListener() {
            @Override
            public boolean onItemLongClick(AdapterView<?> parent, View view, int position, long id) {
                if (position < 5) {
                    row = 0;
                    col=position;
                } else if (position < 8) {
                    row = 1;
                    col=position - 5;
                } else if (position < 13) {
                    row = 2;
                    col=position - 8;
                }
                return false;
            }
        });
        //长按显示快捷菜单
        myListView.setOnCreateContextMenuListener(new View.OnCreateContextMenuListener() {
            @Override
            public void onCreateContextMenu(ContextMenu menu, View v, ContextMenu.ContextMenuInfo menuInfo) {
                menu.setHeaderTitle(getString(title[row][col]));       //设置快捷菜单标题
                menu.add(0, CONTEXT_MENU_00, 0, "立即购买");    //设置快捷菜单项
                menu.add(0, CONTEXT_MENU_01, 1, "加入购物车");
```

```
                    menu.add(0, CONTEXT_MENU_02, 2, "详细信息");
                }
            });
        }
        //点选快捷菜单项要实现的功能
        @Override
        public boolean onContextItemSelected(MenuItem item) {
            switch (item.getItemId()) {
                case CONTEXT_MENU_00:
                    Toast.makeText(BookActivity.this, "您选购了：《" + getString(title[row][col]) , Toast.LENGTH_SHORT).show();
                    return true;
                case CONTEXT_MENU_01:
                    Toast.makeText(BookActivity.this, "已加入购物车，" + getString(info[row][col]), Toast.LENGTH_SHORT).show();
                    return true;
                case CONTEXT_MENU_02:
                    Intent intent = new Intent(this, DetailsActivity.class);
                    Bundle bundle = new Bundle();
                    bundle.putInt("row", row);
                    bundle.putInt("col", col);
                    intent.putExtras(bundle);
                    startActivity(intent);
                    return true;
            }
            return false;
        }
    }
```

3. 设计图书详细信息页

（1）创建 details.xml，设计界面如图 6.6 所示。

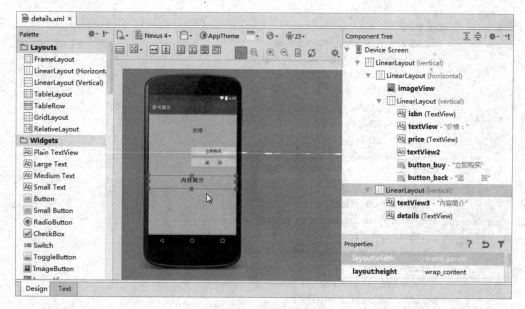

图 6.6　设计详细信息页

图书详细信息页 details.xml 文件代码如下：

```xml
<?xml version="1.0" encoding="utf-8"?>
<LinearLayout xmlns:android="http://schemas.android.com/apk/res/android"
    android:orientation="vertical" android:layout_width="match_parent"
    android:layout_height="match_parent"
    android:background="#aabbcc">

    <LinearLayout
        android:orientation="horizontal"
        android:layout_width="match_parent"
        android:layout_height="260dp">

        <ImageView
            android:layout_width="182dp"
            android:layout_height="256dp"
            android:id="@+id/imageView"
            android:paddingLeft="5dp" />

        <LinearLayout
            android:orientation="vertical"
            android:layout_width="fill_parent"
            android:layout_height="wrap_content">

            <TextView
                android:layout_width="wrap_content"
                android:layout_height="wrap_content"
                android:id="@+id/isbn"
                android:textSize="@dimen/abc_text_size_large_material"
                android:paddingLeft="10dp"
                android:paddingTop="15dp"
                android:textColor="@android:color/black" />

            <TextView
                android:layout_width="wrap_content"
                android:layout_height="wrap_content"
                android:text="价格："
                android:id="@+id/textView"
                android:textColor="@android:color/black"
                android:paddingLeft="10dp"
                android:paddingTop="15dp"
                android:textSize="@dimen/abc_text_size_large_material" />

            <TextView
                android:layout_width="wrap_content"
                android:layout_height="wrap_content"
                android:id="@+id/price"
                android:textColor="@android:color/holo_red_dark"
                android:textSize="@dimen/abc_text_size_display_1_material"
                android:textStyle="bold"
                android:paddingLeft="10dp" />
```

```xml
<TextView
    android:layout_width="fill_parent"
    android:layout_height="10dp"
    android:id="@+id/textView2" />

<Button
    android:layout_width="fill_parent"
    android:layout_height="wrap_content"
    android:text="立即购买"
    android:id="@+id/button_buy"
    android:textColor="@android:color/black"
    android:textSize="@dimen/abc_text_size_medium_material"
    android:onClick="onBuyClick" />                //①

<Button
    android:layout_width="fill_parent"
    android:layout_height="wrap_content"
    android:text="返        回"
    android:id="@+id/button_back"
    android:textColor="@android:color/black"
    android:textSize="@dimen/abc_text_size_medium_material"
    android:onClick="onBackClick" />               //②

    </LinearLayout>
</LinearLayout>

<LinearLayout
    android:orientation="vertical"
    android:layout_width="match_parent"
    android:layout_height="wrap_content">

    <TextView
        android:layout_width="fill_parent"
        android:layout_height="wrap_content"
        android:text="内容简介"
        android:id="@+id/textView3"
        android:textColor="@android:color/black"
        android:textSize="@dimen/abc_text_size_headline_material"
        android:textStyle="bold"
        android:textAlignment="center"
        android:paddingTop="5dp" />

    <TextView
        android:layout_width="wrap_content"
        android:layout_height="wrap_content"
        android:id="@+id/details"
        android:textSize="@dimen/abc_text_size_medium_material"
        android:textStyle="bold"
        android:paddingTop="5dp"
        android:singleLine="false"
        android:maxLines="100"
```

```
            android:scrollbars="vertical"/>
        </LinearLayout>
</LinearLayout>
```

其中：

① 单击"立即购买"按钮，执行 onBuyClick 方法。

② 单击"返回"按钮，执行 onBackClick 方法。

（2）创建 DetailsActivity.java 文件，用于实现详细信息页的功能。

DetailsActivity.java 文件的代码如下：

```
package com.easybooks.android.bookpage;
import …
public class DetailsActivity extends AppCompatActivity {
    private TextView isbnView;
    private TextView priceView;
    private TextView detailsView;
    private ImageView imageView;
    private int row = 0;
    private int col = 0;
```

> 定义图书封面图片(image)、书名(title)、书号(info) 数组

> 定义图书价格(price)、内容简介(details)数组

```
    private int[][] price = {
            {R.string.price_db0, R.string.price_db1, R.string.price_db2, R.string.price_db3, R.string.price_db4},
            {R.string.price_java0, R.string.price_java1, R.string.price_java2},
            {R.string.price_qt0, R.string.price_qt1, R.string.price_qt2, R.string.price_qt3, R.string.price_qt4}
    };                      //价格
    private int[][] details = {
            {R.string.details_db0, R.string.details_db1, R.string.details_db2, R.string.details_db3, R.string.details_db4},
            {R.string.details_java0, R.string.details_java1, R.string.details_java2},
            {R.string.details_qt0, R.string.details_qt1, R.string.details_qt2, R.string.details_qt3, R.string.details_qt4}
    };                      //内容简介

    @Override
    protected void onCreate(Bundle savedInstanceState) {
        super.onCreate(savedInstanceState);
        setContentView(R.layout.details);
        findViews();
        showDetails();
    }
    //设置页面控件引用
    private void findViews() {
        isbnView = (TextView) findViewById(R.id.isbn);
        priceView = (TextView) findViewById(R.id.price);
        detailsView = (TextView) findViewById(R.id.details);
        //给"内容简介"下方的文本控件添加滚动条以显示更多内容
        detailsView.setMovementMethod(ScrollingMovementMethod.getInstance());
        imageView = (ImageView) findViewById(R.id.imageView);
```

```java
    }
    //根据BookActivity.java类显示的页面选择的图书数组二维下标显示图书信息
    private void showDetails() {
        Bundle bundle = getIntent().getExtras();
        row = bundle.getInt("row");
        col = bundle.getInt("col");
        imageView.setImageResource(image[row][col]);
        isbnView.setText(isbn[row][col]);
        priceView.setText(price[row][col]);
        detailsView.setText(details[row][col]);
    }

    public void onBuyClick(View view){
        Toast.makeText(DetailsActivity.this, "您选购了：《" + getString(title[row][col]) + "》，价格：" + getString(price[row][col]), Toast.LENGTH_SHORT).show();
    }

    public void onBackClick(View view){
        finish();
    }
}
```

（3）在AndroidManifest.xml中配置，增加图书详细信息显示DetailsActivity。

```xml
<?xml version="1.0" encoding="utf-8"?>
<manifest xmlns:android="http://schemas.android.com/apk/res/android"
    package="com.easybooks.android.bookpage">

    <application
        ...
        <activity
            android:name=".BookActivity"
            ...
        </activity>
        <activity
            android:label="详细信息"
            android:name=".DetailsActivity"/>
    </application>
</manifest>
```

4. 运行菜单

程序运行效果如图6.7所示。

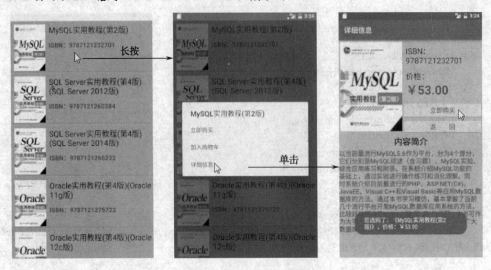

图 6.7 快捷菜单运行效果

6.1.4 操作栏（【例四】：实现例二分组菜单）

操作栏（Action Bar）是 Android 3.0 新引入的界面元素，代替传统的标题栏功能，如图 6.8 所示是电子邮件程序的操作栏。操作栏左侧的图标是应用程序的图标（Logo），图标旁边是应用程序当前 Activity 的标题，右侧的多个图标则是选项菜单中的菜单项。

图 6.8 电子邮件程序的操作栏

操作栏右侧所显示的内容，会根据操作栏所具有的空间不同而具有不同的显示方式。在屏幕尺寸较小的设备上，操作栏会自动隐藏菜单项的文字，而仅显示菜单项的图标。而在屏幕尺寸较大的设备上，操作栏会同时显示菜单项的文字和图标。

默认情况下，基于 holographic 主题的 Activity 上方都存在操作栏。如果需要隐藏 Activity 的操作栏，可以在 AndroidManifest.xml 文件中添加如下代码：

＜activity   android:theme=" @ android: style/Theme .Holo .NoActionBar ">

或者在代码中加入：

ActionBar actionBar= getActionBar ( ) ;
actionBar.hide () ;

在操作栏被隐藏后，Android 系统会自动调整界面元素，填充隐藏操作栏所腾出的空间。

操作栏提供的功能包括下列几个方面。

（1）将选项菜单的菜单项显示在操作栏的右侧。

（2）基于 Fragment 实现类似于 Tab 页的导航切换功能。

（3）为导航提供可拖曳—放置的下拉列表。

（4）可在操作栏上实现类似于搜索框的功能。

实现 ActionBar 与 OptionMenu 代码基本相同，基本思想都是使用 XML 文件的菜单资源，然后在 Activity 中通过 onCreateOptionsMenu()函数加载选项菜单，并调用 onOptionsItemSelected()函数处理菜单选择事件。

不同之处在于 ActionBar 示例 menu_main.xml 文件中的所有菜单项都添加：
```
android:showAsAction="ifRoom|withText"
```
属性修改而来。ifRoom 表示如果操作栏有剩余空间，则显示该菜单项的图标；withText 表示显示图标的同时显示文字标题。

【例四】在子菜单工程 SubMenu（登录和注册为用户组，图书分类显示为图书组）的基础上修改成操作栏的形式。

（1）新建一个文件夹 ActionBar，将工程 SubMenu 复制出来放到该文件夹下，打开工程。

（2）将选项菜单的菜单项标识为可在操作栏中显示的代码非常简单，只需要在 XML 菜单资源文件的 item 标签中添加下列代码：

```
app:showAsAction="ifRoom|withText"
```

menu_main.xml 文件的代码修改如下：

```xml
<?xml version="1.0" encoding="utf-8"?>
<menu xmlns:android="http://schemas.android.com/apk/res/android"
    xmlns:app="http://schemas.android.com/apk/res-auto">
    <item android:id="@+id/sub_menu_0_0"
        android:icon="@drawable/log"
        android:title="登录"
        app:showAsAction="ifRoom|withText"/>
    <item android:id="@+id/sub_menu_0_1"
        android:icon="@drawable/reg"
        android:title="注册"
        app:showAsAction="ifRoom|withText"/>
    <item android:id="@+id/sub_menu_1_0"
        android:icon="@drawable/allicon"
        android:title="所有图书"
        app:showAsAction="ifRoom|withText"/>
    <item android:id="@+id/sub_menu_1_1"
        android:icon="@drawable/dbicon"
        android:title="数据库"
        app:showAsAction="ifRoom|withText"/>
    <item android:id="@+id/sub_menu_1_2"
        android:icon="@drawable/javaicon"
        android:title="Java 开发"
        app:showAsAction="ifRoom|withText"/>
    <item android:id="@+id/sub_menu_1_3"
        android:icon="@drawable/qticon"
        android:title="其他"
        app:showAsAction="ifRoom|withText"/>
</menu>
```

ActionBar 示例的运行界面如图 6.9 所示，其中图 6.9（a）是手机屏幕为纵向时的显示效果，图 6.9（b）是屏幕横向时的显示效果。由此可见，手机屏幕的方向在一定的程度上决定了操作栏的显示内容和显示方式。

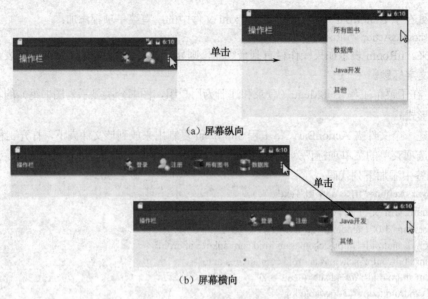

(a) 屏幕纵向

(b) 屏幕横向

图 6.9 操作栏的运行效果

6.2 Fragment 页面（【例五】：图书列表和详细信息不同页和同页显示）

Fragment 是 Android 3.0 新引入的概念，主要目的是在大屏幕设备上实现灵活、动态的界面设计。例如，在 Android 的平板电脑上，因为屏幕有更多的空间来放置更多的界面组件，并且这些组件之间还会产生一定的数据交互。

如果希望将画面分割成左、中、右 3 个部分，效果如图 6.10 所示。

图 6.10 画面分割成左、中、右 3 个部分（Fragment）

通过设计 3 个 Fragment，这 3 个部分不仅画面各自独立，就连操作内容也各自独立，所以开发者需要一种可以粘贴在 Activity 页面上的组件，而且这个组件要有自己独立的 layout 文件、独立的事件处理甚至独立的生命周期。

实际上，Fragment 极具弹性，既可以是 Activity 的一部分，也可以占满整个 Activity 画面。在性质上非常类似于 Activity，可以把它称为 Sub Activity。

通过 Fragment，开发人员无须管理复杂的视图结构变化，而把这些动态的管理工作交给 Fragment 和回退堆栈(back stack)完成。在进行界面设计时，只需要将界面布局按照功能和区域划分为不同的模块，每个模块设计成一个 Fragment 即可。

（1）多个 Fragment 在一个 Activity 中

例如，在新闻阅读程序中，可以将界面划分为左右两部分，左侧用来展示新闻列表，右侧用来阅读新闻的具体内容。如果不使用 Fragment，开发人员就需要在一个 Activity 中实现展示新闻列表，而在另一个 Activity 中显示新闻的具体内容。

使用 Fragment，两个 Fragment 可以并排地放置在同一个 Activity 中，且这两个 Fragment 都具有自己的生命周期回调函数和界面输入事件。如图 6.11 所示。

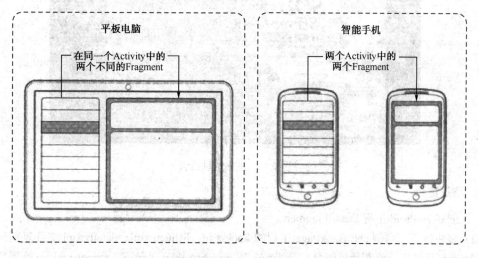

图 6.11　Fragment 与 Activity 的关系

（2）一个 Fragment 在多个 Activity 中

Fragment 是可以被设计成为可重用模块的，这是因为每个 Fragment 都有自己的布局和生命周期回调函数，可以将同一个 Fragment 放置到多个不同的 Activity 中。这样，在设计时为了可重用 Fragment，开发人员应该避免直接从一个 Fragment 去操纵另一个 Fragment，这样就增加了两个 Fragment 之间的耦合度，不利于模块的重用。

（3）Fragment 组合适应不同尺寸的屏幕

通过不同的 Fragment 组合，可以适应不同尺寸的屏幕。例如，对于新闻阅读程序，如果需要同时支持平板电脑和智能手机，则可以重用为平板电脑设计的两个 Fragment，在智能手机端将两个 Fragment 加载到两个 Activity 中。

【例五】采用 Fragment，对于图书列表和详细信息，横向用两个页面图书信息显示，纵向用一个页面图书信息显示。如图 6.12 所示。

图 6.12 图书信息显示

1. 准备资源

（1）创建 Android 工程 DetailFragment。

（2）将快捷菜单实例 ch6\BookPage 工程的 book.xml、listitem.xml、details.xml 文件复制到当前工程 res\layout 目录下；将图片资源复制到当前工程 res\drawable 目录下；strings.xml 替换到当前工程 res\values 目录下的同名文件。

2. 创建布局模式文件

为了使程序能根据智能设备的屏幕宽度自动调整（自适应）横排与竖排两种模式，需要在工程中创建两个用于布局的 XML 文件，并进行相关的设置。

方法如下：

（1）在 res\layout 目录下创建 landscape.xml 和 portrait.xml。

landscape.xml 文件用于双版块模式，代码如下：

```
<?xml version="1.0" encoding="utf-8"?>
<LinearLayout xmlns:android="http://schemas.android.com/apk/res/android"
    android:orientation="horizontal" android:layout_width="match_parent"
    android:layout_height="match_parent">

    <FrameLayout
        android:layout_width="0dp"
        android:layout_height="match_parent"
```

```
            android:id="@+id/master"
            android:layout_weight="1"
            android:background="@android:color/holo_blue_bright">

        </FrameLayout>

        <FrameLayout
            android:layout_width="0dp"
            android:layout_height="match_parent"
            android:id="@+id/detail"
            android:layout_weight="5.2"
            android:background="#aabbcc">

        </FrameLayout>
</LinearLayout>
```

portrait.xml 文件用于单版块模式,代码如下:

```
<?xml version="1.0" encoding="utf-8"?>
<LinearLayout xmlns:android="http://schemas.android.com/apk/res/android"
    android:orientation="vertical" android:layout_width="match_parent"
    android:layout_height="match_parent">

    <FrameLayout
        android:layout_width="match_parent"
        android:layout_height="match_parent"
        android:id="@+id/master"
        android:background="@android:color/holo_blue_bright">

    </FrameLayout>
</LinearLayout>
```

(2)在 res 目录下创建一个名为 values-land 的文件夹,在 values 和 values-land 文件夹下(用 Windows 记事本)新建一个空的 layouts.xml 文件,操作完成回到工程树,展开可见 app\res\values\layouts.xml 下有两个 layouts.xml 文件,如图 6.13 所示。

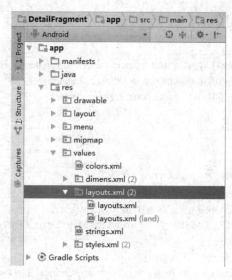

图 6.13 两种模式的布局配置文件

将 layouts.xml(land)内容配置为：

```xml
<?xml version="1.0" encoding="utf-8"?>
<resources>
    <item name="master" type="layout">@layout/landscape</item>
</resources>
```

将 layouts.xml 的内容配置为：

```xml
<?xml version="1.0" encoding="utf-8"?>
<resources>
    <item name="master" type="layout">@layout/portrait</item>
</resources>
```

注意，这两个资源配置 layouts.xml 文件的<item>的 name 属性必须与由 type 属性指定的 xml 文件的第一个 Fragment 的 ID 属性相同。这里均为"master"。

3. 创建 Fragment 类

每一个版块对应创建一个 Fragment 类。纵向 xml 文件双版块模式 Fragment，第一个 Fragment 显示图书列表，列表内容包含图书封面、书名和 ISBN，采用前面实例 book.xml 页面，需要版块对应于 BookFragment 类。第二个 Fragment 用于显示详细信息，采用前面实例 details.xml 页面。一个横向 xml 文件单版块模式 Fragment，用于显示图书详细信息，也采用 details.xml 页面，所以需要版块对应于 DetailsFragment 类。

（1）创建图书版块对应于 BookFragment 类，BookFragment.java 代码如下。

```java
package com.easybooks.android.detailfragment;

import android.app.Activity;
...
public class BookFragment extends Fragment {
    private ListView listView;
    private int row = 0;
    private int col = 0;
    private boolean land = false;
    private static String TAG;

    定义图书封面图片(image)、书名(title)、书号(info) 数组

    @Override
    public View onCreateView(LayoutInflater inflater, ViewGroup container, Bundle savedInstanceState) {
        super.onCreateView(inflater, container, savedInstanceState);
        View view = inflater.inflate(R.layout.book, container, false);
        findViews();
        return view;
    }

    private void findViews() {
        //显示图书列表
        listView = (ListView) view.findViewById(R.id.listView);
        List<Map<String, Object>> list = new ArrayList<Map<String, Object>>();
        for (int i = 0; i < 3; i++) {
            for (int j = 0; j < image[i].length; j++) {
                Map<String, Object> item = new HashMap<String, Object>();
                item.put("image", image[i][j]);
```

```java
                    item.put("title", getResources().getString(title[i][j]));
                    item.put("info", getResources().getString(info[i][j]));
                    list.add(item);
                }
            }
            SimpleAdapter simpleAdapter = new SimpleAdapter(getActivity(), list, R.layout.listitem, new String[]{"image", "title", "info"}, new int[]{R.id.image, R.id.title, R.id.info});
            listView.setAdapter(simpleAdapter);
            //注册监听器,获取用户点选的书目位置，然后转换成列表信息的数组二维下标
            listView.setOnItemClickListener(new AdapterView.OnItemClickListener() {
                @Override
                public void onItemClick(AdapterView<?> parent, View view, int position, long id) {
                    if (position < 5) {
                        row = 0;
                        col=position;
                    } else if (position < 8) {
                        row = 1;
                        col=position - 5;
                    } else if (position < 13) {
                        row = 2;
                        col=position - 8;
                    }
                    land – getArguments().getBoolean("land");
                    //由"图书版块"启动"详情版块"
                    FragmentManager manager = getActivity().getSupportFragmentManager();
                    FragmentTransaction transaction = manager.beginTransaction();
                    DetailsFragment detailsFragment = new DetailsFragment();
                    Bundle bundle = new Bundle();
                    bundle.putInt("row", row);
                    bundle.putInt("col", col);
                    bundle.putBoolean("land", land);
                    detailsFragment.setArguments(bundle);
                    TagString tagString = new TagString();
                    //如果由"图书版块"进入"详情版块"，则先加后剥，否则版块代替
                    if (!land) {
                        transaction.add(R.id.master, detailsFragment, tagString.getTAG_Details());
                        Fragment fragment = manager.findFragmentByTag(tagString.getTAG_Book());
                        transaction.detach(fragment);        //剥离"图书版块"
                    } else
                        transaction.replace(R.id.detail, detailsFragment, tagString.getTAG_Details());
                    transaction.commit();
                }
            });
        }
    }
```

（2）创建详细版块对应于 DetailsFragment 类，DetailsFragment.java 代码如下：

```java
package com.easybooks.android.detailfragment;
import …

public class DetailsFragment extends Fragment {
    private TextView isbnView;
```

```java
            private TextView priceView;
            private TextView detailsView;
            private ImageView imageView;
            private Button mybuy;
            private Button myback;
            private int row = 0;
            private int col = 0;
            private boolean land = false;
```

> 定义图书封面图片(image)、书名(title)、书号(info) 数组

> 定义图书价格(price)、内容简介(details)数组

```java
            @Override
        public View onCreateView(LayoutInflater inflater, ViewGroup container, Bundle savedInstanceState) {
            super.onCreateView(inflater, container, savedInstanceState);
            View view = inflater.inflate(R.layout.details, container, false);
            isbnView = (TextView) view.findViewById(R.id.isbn);
            priceView = (TextView) view.findViewById(R.id.price);
            detailsView = (TextView) view.findViewById(R.id.details);
            //给"内容简介"下方的文本控件添加滚动条以显示更多内容
            detailsView.setMovementMethod(ScrollingMovementMethod.getInstance());
            imageView = (ImageView) view.findViewById(R.id.imageView);
            mybuy = (Button) view.findViewById(R.id.button_buy);
            myback = (Button) view.findViewById(R.id.button_back);
            land = getArguments().getBoolean("land");
            //如果是从图书列表页到当前详细信息页,"返回"按钮不能用、不可见
            if (land) {
                myback.setEnabled(false);
                myback.setVisibility(View.INVISIBLE);
            }
            showDetails();
            //单击"立即购买",显示选择图书信息
            mybuy.setOnClickListener(new View.OnClickListener() {
                @Override
                public void onClick(View v) {
                    Toast.makeText(getActivity(), "您选购了:《" + getString(title[row][col]) + "》,价格:" + getString(price[row][col]), Toast.LENGTH_SHORT).show();
                }
            });
            //监听"返回"按钮
            myback.setOnClickListener(new View.OnClickListener() {
                @Override
                //单击"返回"按钮,移除"详情版块",显示"图书版块"
                public void onClick(View v) {
                    //返回"图书版块"
                    FragmentManager manager = getActivity().getSupportFragmentManager();
                    FragmentTransaction transaction = manager.beginTransaction();
                    TagString tagString = new TagString();
                    Fragment fragment = manager.findFragmentByTag(tagString.getTAG_Book());
```

```
                transaction.attach(fragment);
                //同时移除"详情版块"
                fragment = manager.findFragmentByTag(tagString.getTAG_Details());
                transaction.remove(fragment);
                transaction.commit();
            }
        });
        return view;
    }

    private void showDetails() {
        row = getArguments().getInt("row");
        col = getArguments().getInt("col");
        imageView.setImageResource(image[row][col]);
        isbnView.setText(isbn[row][col]);
        priceView.setText(price[row][col]);
        detailsView.setText(details[row][col]);
    }
}
```

（3）创建 TagString 类用于统一管理各个 Fragment 类的标识，TagString.java 代码如下：

```
package com.easybooks.android.detailfragment;

public class TagString {
    private final static String TAG_Book = "图书版块（BookFragment）";
    private final static String TAG_Details="详情版块（DetailsFragment）";

    public String getTAG_Book() {
        return TAG_Book;
    }

    public String getTAG_Details(){
        return TAG_Details;
    }
}
```

4. 实现主控 Activity

（1）创建 MasterActivity 类，作为本项目程序的主控类实现初始加载和启动版块的功能。MasterActivity.java 代码如下：

```
package com.easybooks.android.detailfragment;
import ...
public class MasterActivity extends AppCompatActivity {
    private boolean land = false;               //是否横排（双窗格）模式

    @Override
    protected void onCreate(Bundle savedInstanceState) {
        super.onCreate(savedInstanceState);
        setContentView(R.layout.master);
        View detailFrame = findViewById(R.id.detail);
        //判断是否在"详情版块"
        if (detailFrame != null) land = true;
```

```
            //加载"图书版块"
            FragmentManager manager = getSupportFragmentManager();
            FragmentTransaction transaction = manager.beginTransaction();
            BookFragment bookFragment = new BookFragment();
            Bundle bundle = new Bundle();
            bundle.putBoolean("land", land);
            bookFragment.setArguments(bundle);
            TagString tagString = new TagString();
            transaction.replace(R.id.master, bookFragment, tagString.getTAG_Book());
            if (land) {            //如果横向,加载"详情版块"
                DetailsFragment detailsFragment = new DetailsFragment();
                bundle.putInt("row", 0);
                bundle.putInt("col", 0);
                detailsFragment.setArguments(bundle);
                transaction.add(R.id.detail, detailsFragment, tagString.getTAG_Details());
            }
            transaction.commit();
        }
    }
```

（2）在 AndroidManifest.xml 中配置主启动的 Activity，如下所示。

```xml
<?xml version="1.0" encoding="utf-8"?>
<manifest xmlns:android="http://schemas.android.com/apk/res/android"
    package="com.easybooks.android.detailfragment">

    <application
     ...
        <activity
            android:name=".MasterActivity"
         ...
        </activity>
    </application>

</manifest>
```

5. 运行版块页

程序运行，过程说明如下。

（1）系统根据 AndroidManifest.xml 文件，找到 android:name=".MasterActivity"属性，然后执行 MasterActivity.java 类代码。

（2）在 MasterActivity.java 类中根据 onCreate(…) 中的"setContentView(R.layout.master);"代码，确定 master 资源名页面显示，根据手机（或者模拟器）当前是纵向还是横向，如果横向，找包含两个 Fragment，name="master"的页面显示，否则找包含一个 Fragment，name="master"的页面显示。

先显示 BookFragment 图书列表页面，如果是横排，用横排显示 DetailsFragment 单版详细信息。在 BookFragment 图书列表页面，单击列表行，显示纵向第二个 Fragment 图书详细信息。

显示如图 6.12 所示。

6.3 Tab 导航栏

6.3.1 Tab 导航栏介绍

TabHost 组件可以在界面中存放多个选项卡，其中 TabWidget 表示 TabHost 标签页中上部或者下部的按钮，可以点击按钮切换选项卡。TabSpec 代表了选项卡界面，添加一个 TabSpec 即可添加到 TabHost 中。

1. TabHost 布局文件

在 XML 文件中使用 TabHost 组件，并在其中定义一个 FrameLayout 选项卡内容。可以使用标签设置，其中的 id 需要引用 android 的自带 id：

android:id=@android:id/tabhost。

例如：

```
<tabhost android:id="@android:id/tabhost"
    android:layout_height="match_parent"
    android:layout_width="match_parent"
</tabhost>
```

2. TabWidget 组件

布局文件中 TabWidget 代表的就是选项卡按钮，Fragment 组件代表内容。如果 Fragment 组件没有设置 android：layout_weight 属性，那么将 TabWidget 放到下面，可能不会显示按钮。设置了 Fragment 组件的权重之后，就可以成功显示该选项卡按钮。

例如：

```
<tabwidget android:id="@android:id/tabs"
android:layout_height="wrap_content"
android:layout_width="fill_parent"
android:orientation="horizontal"
</tabwidget>
```

其中引用 android 的自带 id。

3. FrameLayout 组件

FrameLayout 组件是 TabHost 组件中的必备组件，如果想要将按钮放到下面，可以将该组件定义在下面。但是要注意，FrameLayout 要设置 android:layout_widget=1。如果想要设置该按钮组件的大小，可以设置该组件与 FrameLayout 组件的权重。

该组件中定义的子组件是 TabHost 中每个页面显示的选项卡，可以将 TabHost 选项卡显示的视图定义在其中。组件的 id 要设置成 android 的自带的 id。

android:id=@android:id/tabcontent；

示例：

```
<framelayout android:id="@android:id/tabcontent"
android:layout_height="fill_parent"
android:layout_weight="1"
android:layout_width="fill_parent">
</framelayout>
```

4. Activity 方法

setup()：组织 TabHost 组件。

getHost()：获取 TabHost 组件的方法的前提是在布局文件中。但要求设置了 android 自带的 id android:id=@android:id/tabhost。

newTabHost(tag)：创建选项卡。其中的 tag 是字符串，即在选项卡的唯一标识。

setIndicator(x)：设置按钮名称。

setContent()：设置选项卡内容。可以设置视图组件、Activity 和 Fragment。

tabHost.add(tag)：添加选项卡。传入的参数是创建选项卡的时候定义的唯一标识。

setCurrentTabTag(x)：设置 x 为当前选项卡。

getCurrentTabTag(x)：获得当前选项卡标识。

6.3.2 Tab 导航栏应用（【例六】：实现例二分组菜单）

【例六】在子菜单工程 SubMenu（登录和注册为用户组，图书分类显示为图书组）的基础上修改成 Tab 导航栏实现形式。

1. 准备资源

创建 Android 工程 TabHost，准备以下资源：把例二工程 SubMenu 中的 book.xml、listitem.xml 复制到 res\layout 目录下。把图片资源包括图书的子菜单图标 allicon.png、dbicon.png、javaicon.png、qticon.png 复制到 res\drawable 目录下。用 strings.xml 代替当前的同名文件。

2. 设计页面

在 content_main.xml 页面上拖曳一个 TabHost，如图 6.14 所示。

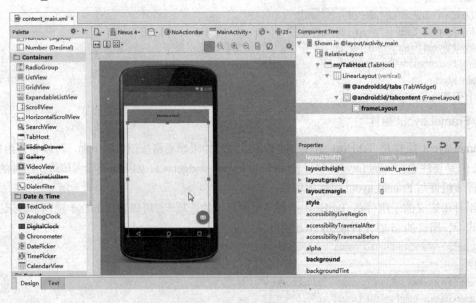

图 6.14 设计视图

content_main.xml 源码如下：

```
<?xml version="1.0" encoding="utf-8"?>
<RelativeLayout xmlns:android="http://schemas.android.com/apk/res/android"
    ...
```

```xml
        android:background="@color/background_material_light">

    <TabHost
        android:layout_width="match_parent"
        android:layout_height="match_parent"
        android:id="@+id/myTabHost"
        android:layout_alignParentTop="true"
        android:layout_centerHorizontal="true">

        <LinearLayout
            android:layout_width="match_parent"
            android:layout_height="match_parent"
            android:orientation="vertical">

            <TabWidget
                android:id="@android:id/tabs"
                android:layout_width="match_parent"
                android:layout_height="wrap_content"
                android:orientation="horizontal"
                android:background="@android:color/holo_blue_light"></TabWidget>

            <FrameLayout
                android:id="@android:id/tabcontent"
                android:layout_width="match_parent"
                android:layout_height="match_parent">

                <FrameLayout
                    android:layout_width="match_parent"
                    android:layout_height="match_parent"
                    android:id="@+id/frameLayout">
                </FrameLayout>
            </FrameLayout>
        </LinearLayout>
    </TabHost>
</RelativeLayout>
```

3. 创建 Fragment 类

对应图书版块创建 BookFragment 类,BookFragment.java 代码如下:

```java
package com.easybooks.android.tabhost;
import ...

public class BookFragment extends Fragment{
    private ListView listView;
    private int[][] image = {
            {R.drawable.db0, R.drawable.db1, R.drawable.db2, R.drawable.db3, R.drawable.db4},
            {R.drawable.java0, R.drawable.java1, R.drawable.java2},
            {R.drawable.qt0, R.drawable.qt1, R.drawable.qt2, R.drawable.qt3, R.drawable.qt4}
    };                          //图书封面图片
    private int[][] title = {
            {R.string.title_db0, R.string.title_db1, R.string.title_db2, R.string.title_db3, R.string.title_db4},
            {R.string.title_java0, R.string.title_java1, R.string.title_java2},
```

```
            {R.string.title_qt0, R.string.title_qt1, R.string.title_qt2, R.string.title_qt3, R.string.title_qt4}
    };                              //书名
    private int[][] info = {
            {R.string.info_db0, R.string.info_db1, R.string.info_db2, R.string.info_db3, R.string.info_db4},
            {R.string.info_java0, R.string.info_java1, R.string.info_java2},
            {R.string.info_qt0, R.string.info_qt1, R.string.info_qt2, R.string.info_qt3, R.string.info_qt4}
    };                              //书号
    private int curgroup;        //当前要预览的图书类别(-1:所有;0:数据库;1:Java 开发;2:其他)

    @Override
    public View onCreateView(LayoutInflater inflater, ViewGroup container, Bundle savedInstanceState) {
        super.onCreateView(inflater, container, savedInstanceState);
        View view = inflater.inflate(R.layout.book, container, false);
        listView = (ListView) view.findViewById(R.id.listView);
        findViews();
        return view;
    }

    private void findViews() {
        curgroup = getArguments().getInt("group");  //获得图书分类号
        List<Map<String, Object>> list = new ArrayList<Map<String, Object>>();
        //按照图书分类号显示图书信息
      for (int i = 0; i < 3; i++) {
            if ((curgroup != -1 && curgroup == i) || curgroup == -1) {
                for (int j = 0; j < image[i].length; j++) {
                    Map<String, Object> item = new HashMap<String, Object>();
                    item.put("image", image[i][j]);
                    item.put("title", getResources().getString(title[i][j]));
                    item.put("info", getResources().getString(info[i][j]));
                    list.add(item);
                }
            }
        }
        SimpleAdapter simpleAdapter = new SimpleAdapter(getActivity(), list, R.layout.listitem, new
String[]{"image", "title", "info"}, new int[]{R.id.image, R.id.title, R.id.info});
        listView.setAdapter(simpleAdapter);
    }
}
```

4. 主程序功能

在 MainActivity.java 中编写代码,实现初始加载和启动版块功能。

(1) 在 findViews()中的功能。

① 组织 tabhost:myTabHost.setup();

② 添加选项卡:myTabHost.addTab(myTabHost.newTabSpec("tab_db").setIndicator("", getResources().getDrawable(R.drawable.dbicon)).setContent(R.id.frameLayout));

包括选项卡标识:tab_db;

采用的图标:drawable\dbicon 文件。

显示的内容:在 content_main.xml 文件<frameLayout>标签指定。

③ 指定当前显示选项卡为第三个(所有):myTabHost.setCurrentTab(3);

显示所有图书信息：previewBook(-1);

（2）previewBook(int group) { ... }

把图书分类号信息传递给 BookFragment 类，在 BookFragment 类定义 findViews()方法显示。

（3）监听选项卡改变 onTabChanged 事件。

setOnTabChangedListener(...){...onTabChanged(){...}}

根据选择的选项卡，显示相应的图书信息。

MainActivity.java 代码如下：

```java
package com.easybooks.android.tabhost;
import ...

public class MainActivity extends AppCompatActivity {
    private TabHost myTabHost;

    @Override
    protected void onCreate(Bundle savedInstanceState) {
        super.onCreate(savedInstanceState);
        setContentView(R.layout.activity_main);
        findViews();
        Toolbar toolbar = (Toolbar) findViewById(R.id.toolbar);
        ...
        //监听选项卡改变 onTabChanged 事件
        myTabHost.setOnTabChangedListener(new TabHost.OnTabChangeListener() {
            @Override
            public void onTabChanged(String tabId) {              //选项卡改变 onTabChanged 事件产生
                switch (myTabHost.getCurrentTabTag()) {           //获得当前选项卡标识
                    case "tab_db":                                 //根据选项卡标识，变成图书分类代号
                        previewBook(0);                            //按照图书分类代号显示
                        break;
                    case "tab_java":
                        previewBook(1);
                        break;
                    case "tab_qt":
                        previewBook(2);
                        break;
                    case "tab_all":
                        previewBook(-1);
                        break;
                    default:
                        break;
                }
            }
        });
    }

    private void findViews() {
        myTabHost = (TabHost) findViewById(R.id.myTabHost);
        myTabHost.setup();
        myTabHost.addTab(myTabHost.newTabSpec("tab_db").setIndicator("",
getResources().getDrawable(R.drawable.dbicon)).setContent(R.id.frameLayout));
```

```
            myTabHost.addTab(myTabHost.newTabSpec("tab_java").setIndicator("",
getResources().getDrawable(R.drawable.javaicon)).setContent(R.id.frameLayout));
            myTabHost.addTab(myTabHost.newTabSpec("tab_qt").setIndicator("",
getResources().getDrawable(R.drawable.qticon)).setContent(R.id.frameLayout));
            myTabHost.addTab(myTabHost.newTabSpec("tab_all").setIndicator("所有图书").setContent(R.id.
frameLayout));
            myTabHost.setCurrentTab(3);
            previewBook(-1);
        }

        public void previewBook(int group) {
            //根据用户点击 Tab 选择的类别,向"图书版块"传递 group 参数以显示不同类别书的预览
            FragmentManager manager = getSupportFragmentManager();
            FragmentTransaction transaction = manager.beginTransaction();
            BookFragment bookFragment = new BookFragment();
            Bundle bundle = new Bundle();
            bundle.putInt("group", group);
            bookFragment.setArguments(bundle);
            Fragment fragment = manager.findFragmentById(R.id.frameLayout);
            if (fragment == null) transaction.add(R.id.frameLayout, bookFragment);
            else transaction.replace(R.id.frameLayout, bookFragment);
            transaction.commit();
        }
        ...
    }
```

5. 运行导航栏

程序运行,点击导航栏上不同的导航图标,列表显示不同类别的图书信息,如图 6.15 所示。

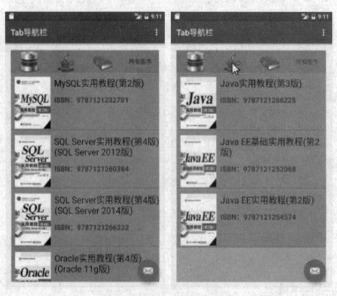

图 6.15 导航栏的运行效果

第 7 章　Android 服务与广播程序设计

7.1　Java 线程编程基础

线程是独立的程序单元，多个线程可以并行工作。在多处理器系统中，每个中央处理器(CPU)单独运行一个线程，因此线程是并行工作的。但在单处理器系统中，处理器会给每个线程一小段时间，在这个时间内执行该线程，然后处理器执行下一个线程，这样也可以认为线程并行运行。无论线程是否真的并行工作，在宏观上可以认为子线程是独立于主线程，且能与主线程并行工作的程序单元。

在 Android 系统中需要复杂运算过程、大量的文件操作、存在延时的网络通信和数据库操作等耗时的处理过程。如果它们工作在 Activity、Service 和 BroadcastReceiver 等主线程上，就会降低用户界面的响应速度，甚至导致用户界面失去响应。当用户界面失去响应超过 5 秒后，Android 系统会允许用户强行关闭应用程序。

较好的解决方法是将耗时的处理过程转移到子线程上，这样可以缩短主线程的事件处理时间，从而避免用户界面长时间失去响应。

1. 线程使用

在 Java 语言中，建立和使用线程的方法如下。

（1）需要实现 Java 的 Runnable 接口，并重载 run()函数，在 run()中放置代码的主体部分。

```
private Runnable backgroundWork = new Runnable(){
    @Override
    public void run() {
    //过程代码
    }
};
```

（2）创建线程（Thread）对象，并将 Runnable 对象作为参数传递给 Thread 对象。

在 Thread 的构造函数中，第一个参数用来表示线程组，第二个参数是需要执行的 Runnable 对象，第三个参数是线程的名称。

```
private Thread   myThread;
myThread=new Thread(null, backgroundWork, "myThread");
```

（3）调用 start()方法启动线程。

```
myThread.start();
```

（4）线程的终止

① 当线程在 run()方法返回后，线程就自动终止。

② 调用 stop()在外部终止线程。但不推荐使用这种方法，因为该方法并不安全，有可能产生异常。

③ 调用 myThread.interrupt()方法通知线程准备终止，此后 myThread.interrupted()为真。

一般情况下，子线程需要无限运行，所以通常会将程序主体放置在 while()函数内，并调用 Thread.interrupted()方法判断线程是否应被中断。

例如：

```java
public void run() {
    try {
        while(!myThread.interrupted()){
            //过程代码
            myThread.sleep(1000);
        }
    } catch (InterruptedException e) {
        e.printStackTrace();
    }
}
```

注意，myThread.sleep(1000)代码使线程休眠 1000 毫秒。当线程在休眠过程中线程被中断，则会产生 InterruptedException 异常。在中断的线程上调用 sleep()方法，同样会产生 InterruptedException 异常。因此需要捕获 InterruptedException 异常，保证安全终止线程。

2. 线程中数据更新用户界面

Android 系统提供了多种方法解决线程中数据更新用户界面问题，这里仅介绍使用 Handler 更新用户界面的方法。

Handler 允许将 Runnable 对象发送到线程的消息队列中，每个 Handler 实例绑定到一个单独的线程和消息队列上。当用户建立一个新的 Handler 实例时，通过 post()方法将 Runnable 对象从后台线程发送给 GUI 线程的消息队列。当 Runnable 对象通过消息队列后，这个 Runnable 对象将被运行。

例如：
```java
private static Handler myHandler = new Handler();                    //(1)
public static void UpdateGUI(double refreshDouble){                  //(2)
    myHandler.post(refreshLabel);
}
private static Runnable myRunnable = new Runnable(){
    @Override
    public void run() {                                              //(3)
        //过程代码
    }
};
```

其中：

（1）建立了一个静态的 Handler 实例，但这个实例是私有的，因此外部代码并不能直接调用这个 Handler 实例。

（2）UpdateGUI()是公有的界面更新函数，后台线程通过调用该函数，将后台产生的数据 refreshDouble 传递到 UpdateGUI()函数内部，然后直接调用 post()方法，将创建的 Runnable 对象传递给界面线程（主线程）的消息队列中。

（3）Runnable 对象中需要重载的 run()函数，界面更新代码就在这里。

7.2 Service（服务）程序设计

7.2.1 Service 概述

因为手机硬件性能和屏幕尺寸的限制，通常 Android 系统仅允许一个应用程序处于激活状态并显示在手机屏幕上。因此，Android 系统需要一种后台服务机制，允许在没有用户界面的情况下，使程序能够长时间在后台运行,实现应用程序的后台服务功能,并能够处理事件或数据更新。例如音乐播放器软件需要在关闭播放器界面后仍能够保持音乐持续播放。Android 系统提供的 Service(服务)组件不直接与用户进行交互,能够长期在后台运行。

1. Service 的基本概念

Service 是一个没有用户界面、在后台运行的应用组件,相当于后台运行的 Activity。因为 Service 没有用户界面,更加有利于降低系统资源的消耗,而且 Service 比 Activity 具有更高的优先级,因此在系统资源紧张时,Service 不会被 Android 系统优先终止。即使 Service 被系统终止,在系统资源恢复后 Service 也将自动恢复运行状态,因此可以认为 Service 是在系统中永久运行的组件。Service 除了可以实现后台服务功能,还可以用于进程间通信(IPC 机制),解决不同 Android 应用程序进程之间的调用和通信问题。

2. Service 的回调方法

要创建一个 Service,需要创建一个 Java 类,扩展 Service 基类或者它的子类。Service 基类定义了各种回调方法,如表 7.1 所示。

表 7.1 Service 的回调方法

回调方法	说 明
onStartCommand()	当另一个组件(例如 Activity)调用 startService()方法请求服务启动时,调用该方法。如果开发人员实现该方法,则需要在任务完成时调用 stopSelf()或 stopService()方法停止服务。如果仅想提供绑定,则不必实现该方法
onBind()	当其他组件通过调用 bindService()绑定服务时调用这个方法。在该方法的实现中,必须提供客户端与服务通信的接口。该方法必须实现,但是如果不允许绑定,则返回 null
onCreate()	当服务第一次创建时,系统调用该方法执行一次性建立过程(在系统调用 onStartCommand()或 onBind()方法前)。如果服务已经运行,该方法不被调用
onDestroy()	当服务不再使用并将被销毁时,系统调用这个方法。服务应该实现该方法用于清理,如线程注册的侦听器、接收器等资源

3. Service 的创建与注册

创建一个 Android 工程,在该工程中创建一个继承自 Service 类的子类 LocalService,该类有一个抽象方法 onBind(),必须在子类中实现。

Service 需要在 AndroidManifest.xml 文件中进行注册,注册类 LocalService 的代码如下所示。
`<service android: name=".LocalService"/>`
该代码应位于 application 节点内,与 activity 组件的注册位于同一层次。

4. Service 的启动流程

Service 有"启动"(started)和"绑定"(Bound)两种状态。

(1)通过 startService()启动的服务处于"启动的"状态,一旦启动,Service 就在后台运行,即使启动它的应用组件已经被销毁了。通常 started 状态的 Service 执行单任务并且不返回任何结果给启动者。例如下载或上传一个文件,当这项操作完成时,Service 应该停止它本身。在启动方式中,启动 Service 的组件不能够获取 Service 的对象实例,因此无法调用 Service 中的任何函数,也不能够获取 Service 中的任何状态和数据信息。能够以启动方式使用的 Service,需要具备自管理的能力,而且不需要通过函数调用获取 Service 的功能和数据。

（2）通过调用 bindService()来启动的服务处于"绑定"状态。一个绑定的 Service 提供一个允许组件与 Service 交互的接口，可以发送请求，获取返回结果，还可以通过跨进程通信（IPC）来交互。绑定的 Service 只有当应用组件绑定后才能运行，多个组件可以绑定一个 Service，当调用 unbind()方法时，这个 Service 就会被销毁了。在绑定方式中，Service 的使用是通过服务链接(Connection)实现的，服务链接能够获取 Service 的对象实例，因此绑定 Service 的组件可以调用 Service 中实现的函数，或直接获取 Service 中的状态和数据信息。使用 Service 的组件通过 Context.bindService()建立服务链接，通过 Context.unbindService()停止服务链接。如果在绑定过程中 Service 没有启动，Context.bindService()会自动启动 Service，而且同一个 Service 可以绑定多个服务链接，这样可以同时为多个不同的组件提供服务。

当然，这两种使用方法并不是完全独立的，某些情况下可以混合使用启动方式和绑定方式。还是以 MP3 播放器为例，在后台工作的 Service 通过 Context.startService()启动某个音乐播放，但在播放过程中如果用户需要暂停音乐播放，则要通过 Context.bindService()获取服务链接和 Service 对象实例，进而通过调用 Service 对象实例中的函数，暂停音乐播放过程，并保存相关的信息。在这种情况下，如果调用 Context.stopService()并不能停止 Service，需要在所有的服务链接关闭后，Service 才能够真正停止。

startService()的启动流程如图 7.1 所示，其过程描述如下。

说明：

（1）如果 Service 还没有运行，则先调用 onCreate()，然后调用 onStart()。如果 Service 已经运行，则只调用 onStart()，所以一个 Service 的 onStart 方法可能会重复调用多次。

（2）如果直接退出而没有调用 stopService()，Service 会一直在后台运行，该 Service 的调用者在启动后可以通过 stopService 关闭 Service。

当程序使用 startService()和 stopService()启动、关闭服务时，服务与调用者之间基本不存在太多的关联，也无法与访问者进行通信、数据交互等。当服务需要与调用者进行访问调用和数据交互时，应该使用 bindService()和 unbindService()启动、关闭服务。

Context.bindService()的启动流程如图 7.2 所示，其过程描述如下。

（1）onBind()将返回给客户端一个 IBind 接口实例，IBind 允许客户端回调服务的方法，例如得到 Service 的实例、运行状态或其他操作。这时调用者（例如 Activity）会和 Service 绑定在一起，Context 退出了，Service 就会调用 onUnbind→onDestroy 相应退出。

（2）在 Service 每一次的开启关闭过程中，只有 onStart 可被多次调用（通过多次 startService 调用），其他 onCreate、onBind、onUnbind、onDestory 在一个生命周期中只能被调用一次。

（3）如果只是想要启动一个程序，在后台服务长期进行某项任务，那么使用 startService 便可以了。如果想要与正在运行的 Service 取得联系，有两种方法，一种是使用 broadcast，另一种是使用 bindService。前者的缺点是如果交流较为频繁，容易造成性能上的问题，并且 BroadcastReceiver 本身执行代码的时间是很短的，而后者则没有这些问题。因此选择使用 bindService，也同时在使用 startServic 和 bindService 了，这在 Activity 中更新 Service 的某些运行状态是有用的。另外如果服务只是公开一个远程接口，供连接上的客户端远程调用执行方法，此时可以不让服务一开始就运行，而只用 bindService，这样在第一次 bindService 的时候才会创建服务的实例运行它，这会节约很多系统资源。

第 7 章　Android 服务与广播程序设计

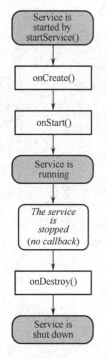

图 7.1　Context.startService()的启动流程

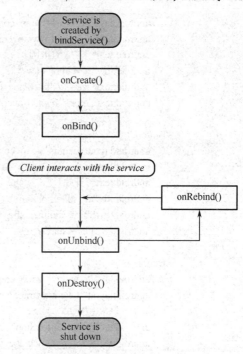

图 7.2　Context.bindService()的启动流程

7.2.2　启动方式使用 Service

【例 7.1】本例演示用启动方式使用服务的过程，在服务运行过程中，主程序无法直接（只能通过句柄提交线程 Runnable 接口）与服务进行交互，两者是相对独立的。

1. 设计页面

创建 Android 工程 CircleStartService。设计页面如图 7.3 所示。

界面上放置"开始"和"停止"两个按钮，用户通过它们启停服务，中央区的 FrameLayout 用于显示后台服务所绘出的圆。

content_main.xml 文件代码如下：
```
<?xml version="1.0" encoding="utf-8"?>
<RelativeLayout
xmlns:android="http://schemas.android.com/apk/res/android"
    …>

    <LinearLayout
        android:orientation="vertical"
        android:layout_width="match_parent"
        android:layout_height="match_parent">

        <LinearLayout
            android:orientation="horizontal"
            android:layout_width="match_parent"
            android:layout_height="wrap_content">

            <TextView
```

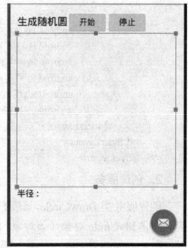

图 7.3　设计视图

```xml
                android:layout_width="wrap_content"
                android:layout_height="wrap_content"
                android:text="生成随机圆"
                android:id="@+id/textView"
                android:textSize="@dimen/abc_text_size_large_material"
                android:textColor="@android:color/black" />

            <Button
                android:layout_width="wrap_content"
                android:layout_height="wrap_content"
                android:text="开始"
                android:id="@+id/buttonStart"
                android:textSize="@dimen/abc_text_size_medium_material"
                android:onClick="start" />

            <Button
                android:layout_width="wrap_content"
                android:layout_height="wrap_content"
                android:text="停止"
                android:id="@+id/buttonStop"
                android:textSize="@dimen/abc_text_size_medium_material"
                android:onClick="stop" />
        </LinearLayout>

        <FrameLayout
            android:layout_width="match_parent"
            android:layout_height="330dp"
            android:id="@+id/circleFrame"></FrameLayout>

        <LinearLayout
            android:orientation="horizontal"
            android:layout_width="match_parent"
            android:layout_height="match_parent">

            <TextView
                android:layout_width="wrap_content"
                android:layout_height="wrap_content"
                android:text="半径："
                android:id="@+id/textRadius"
                android:textSize="@dimen/abc_text_size_medium_material"
                android:textColor="@android:color/black" />
        </LinearLayout>
    </LinearLayout>
</RelativeLayout>
```

2. 创建服务

创建服务类 DrawCircle，该服务以线程的形式在后台运行。先执行 onCreate 方法创建线程对象，接着创建 MyCircle 对象（该对象的 onDraw 方法完成画圆功能）；接着用 onStartCommand 方法启动线程。在 Runnable 的 run 方法中，调用主程序 MainActivity 的 updateGUI 方法刷新界面。

DrawCircle.java 代码如下：

```java
package com.easybooks.android.circlestartservice;
import android.app.Service;
…

public class DrawCircle extends Service {
    private Thread drawThread;                                  //声明线程对象
    private View circle;                                        //声明 MyCircle 对象
    private int radius;                                         //圆半径
    private static final String TAG = "绘圆服务（DrawCircle）";   //标识该服务

    @Override
    public IBinder onBind(Intent intent) {                      //服务必须实现的方法
        return null;
    }

    private Runnable drawWork = new Runnable() {                //(1)
        @Override
        public void run() {
            try {
                while (!drawThread.isInterrupted()) {
                    radius = (int) (Math.random() * 400);
                    MainActivity.updateGUI(circle, radius);
                    drawThread.sleep(1000);
                }
            } catch (InterruptedException e) {
                e.printStackTrace();
            }
        }
    };

    @Override
    public void onCreate() {                                    //(2)
        super.onCreate();
        drawThread = new Thread(null, drawWork, "DrawCircle");
        circle = new MyCircle(this);
        Log.i(TAG, "onCreate");
    }

    @Override
    public int onStartCommand(Intent intent, int flags, int startId) {  //启动服务
        super.onStartCommand(intent, flags, startId);
        if (!drawThread.isAlive()) drawThread.start();
        Log.i(TAG, "onStartCommand");
        return START_STICKY;
    }

    @Override
    public void onDestroy() {                                   //销毁服务
        super.onDestroy();
        drawThread.interrupt();
        Log.i(TAG, "onDestroy");
```

```
    }
    public class MyCircle extends View {                              //(3)
        public MyCircle(Context context) {
            super(context);
        }

        @Override
        protected void onDraw(Canvas canvas) {
            super.onDraw(canvas);
            Paint paint = new Paint();
            canvas.drawCircle(400, 400, radius, paint);
        }
    }
}
```

其中：

（1）"drawWork = new Runnable()" 创建线程运行时要执行的方法，通过重写该接口的 run 方法实现，在 while 循环中，线程不断地随机生成 0~400 之间数值的圆半径，调用主程序 updateGUI 刷新界面，每刷新一次，线程会休眠 1 000 毫秒后继续。

（2）onCreate 是服务生命周期开始执行的第一个方法，在其中创建服务线程及绘圆类对象。

（3）绘圆类 MyCircle 继承自 View 类，通过重写其 onDraw 方法来实现自定义画圆的功能，画圆通过调用 Canvas（画布）对象的 drawCircle 方法实现。

3. 主程序启动

在主程序的 updateGUI 方法中通过 Handler（句柄）对象的 post 方法提交界面刷新机制，而界面刷新的具体操作在 Runnable 接口 refreshWork 的 run 方法中执行。

MainActivity.java 代码如下：

```
package com.easybooks.android.circlestartservice;
import ...
public class MainActivity extends AppCompatActivity {
    private static FrameLayout circleFrame;
    private static TextView textRadius;
    private static Handler handler = new Handler();        //界面刷新句柄
    private static View mycircle;                          //圆视图对象
    private static int myradius;                           //圆半径

    @Override
    protected void onCreate(Bundle savedInstanceState) {
        super.onCreate(savedInstanceState);
        setContentView(R.layout.activity_main);
        findViews();
        Toolbar toolbar = (Toolbar) findViewById(R.id.toolbar);
        ...
    }

    private void findViews() {
        circleFrame = (FrameLayout) findViewById(R.id.circleFrame);
        textRadius = (TextView) findViewById(R.id.textRadius);
    }
```

```java
        public static void updateGUI(View c, int r) {                    //(1)
            mycircle = c;
            myradius = r;
            handler.post(refreshWork);
        }

        private static Runnable refreshWork = new Runnable() {           //(2)
            @Override
            public void run() {
                //刷新界面
                circleFrame.removeAllViews();
                circleFrame.addView(mycircle);
                textRadius.setText("半径: " + String.valueOf(myradius));
            }
        };

        //开启服务
        public void start(View view) {
            Intent intent = new Intent(this, DrawCircle.class);
            startService(intent);
        }

        //停止服务
        public void stop(View view) {
            Intent intent = new Intent(this, DrawCircle.class);
            stopService(intent);
        }
        …
    }
```

其中：

（1）静态方法 updateGUI 负责刷新界面，它主要完成两个工作：一是获取服务中的圆视图对象及半径值；二是通过 Handler 提交到线程 Runnable 接口中。

（2）在线程 Runnable 接口中完成刷新的任务，要做以下三件事：一是用 FrameLayout 的 removeAllViews 方法清除界面上旧有的圆；二是用 addView 方法添加线程新绘制的圆；三是用 setText 方法在界面上显示圆的半径。

4. 服务配置

在 AndroidManifest.xml 中，将开发的服务（DrawCircle）配置其中（加黑处）：

```xml
<?xml version="1.0" encoding="utf-8"?>
<manifest xmlns:android="http://schemas.android.com/apk/res/android"
    package="com.easybooks.android.circlestartservice">

    <application
        …
        <activity
            android:name=".MainActivity"
        …
        </activity>
```

```
<service android:name=".DrawCircle"/>
    </application>

</manifest>
```

5. 运行效果

运行程序，点击"开始"，启动服务线程，在屏幕中央每隔 1 000 毫秒会出现一个不同大小的圆，左下角对应显示该圆的半径值。点击"停止"关闭服务，绘圆的过程就此终止，界面停住不再有变化，效果如图 7.4 所示。

在开发环境底部的日志窗口还可观察到服务启停整个生命周期中所调用的方法，如图 7.5 所示。

```
12-08 06:18:39.071 25144-25144/com.easybooks.android.circlestartservice I/绘圆服务（DrawCircle）: onCreate
12-08 06:18:39.073 25144-25144/com.easybooks.android.circlestartservice I/绘圆服务（DrawCircle）: onStartCommand
12-08 06:23:40.686 25144-25144/com.easybooks.android.circlestartservice I/绘圆服务（DrawCircle）: onDestroy
```

图 7.4　启动绘圆服务运行效果　　　　　　　图 7.5　启动服务的生命周期

7.2.3　绑定方式使用 Service

【例 7.2】本例演示用绑定方式使用服务的过程，与启动方式最大的不同在于：主程序可随时直接主动去获取服务中的数据，无须依赖于 Runnable 接口，此方式适用于主应用程序与服务需要频繁交互和共享数据的场合。

1. 设计页面

创建 Android 工程 CircleBindService，设计页面如图 7.6 所示，与上例的界面布局基本相同，只是在底部增加了一个"计算"按钮及显示圆周长和面积的 TextView。

content_main.xml 文件代码如下：

```xml
<?xml version="1.0" encoding="utf-8"?>
<RelativeLayout       xmlns:android="http://schemas.android.com/apk/res/android"
    …>

    <LinearLayout
        android:orientation="vertical"
        android:layout_width="match_parent"
        android:layout_height="match_parent">

        <LinearLayout
            android:orientation="horizontal"
            android:layout_width="match_parent"
            android:layout_height="wrap_content">

            <TextView
                android:layout_width="wrap_content"
                android:layout_height="wrap_content"
                android:text="生成随机圆"
                android:id="@+id/textView"
                android:textSize="@dimen/abc_text_size_large_material"
```

图 7.6　设计视图

```xml
            android:textColor="@android:color/black" />

        <Button
            android:layout_width="wrap_content"
            android:layout_height="wrap_content"
            android:text="开始"
            android:id="@+id/buttonStart"
            android:textSize="@dimen/abc_text_size_medium_material"
            android:onClick="bind" />

        <Button
            android:layout_width="wrap_content"
            android:layout_height="wrap_content"
            android:text="停止"
            android:id="@+id/buttonStop"
            android:textSize="@dimen/abc_text_size_medium_material"
            android:onClick="unbind" />
</LinearLayout>

<FrameLayout
    android:layout_width="match_parent"
    android:layout_height="330dp"
    android:id="@+id/circleFrame"></FrameLayout>

<LinearLayout
    android:orientation="horizontal"
    android:layout_width="match_parent"
    android:layout_height="wrap_content">

    <TextView
        android:layout_width="wrap_content"
        android:layout_height="wrap_content"
        android:text="半径："
        android:id="@+id/textRadius"
        android:textSize="@dimen/abc_text_size_medium_material"
        android:textColor="@android:color/black" />

    <Button
        android:layout_width="wrap_content"
        android:layout_height="wrap_content"
        android:text="计算"
        android:id="@+id/buttonCalculate"
        android:textSize="@dimen/abc_text_size_medium_material"
        android:onClick="getCircleParam" />
</LinearLayout>

<LinearLayout
    android:orientation="horizontal"
    android:layout_width="match_parent"
    android:layout_height="match_parent">
```

```xml
            <TextView
                android:layout_width="wrap_content"
                android:layout_height="wrap_content"
                android:text="圆周长：面积："
                android:id="@+id/textCircleParam"
                android:textSize="@dimen/abc_text_size_medium_material" />
        </LinearLayout>
    </LinearLayout>
</RelativeLayout>
```

2. 创建服务

创建服务类 DrawCircle，该服务以线程的形式在后台运行。先执行 onCreate 方法创建线程对象，接着创建 MyCircle 对象（该对象的 onDraw 方法完成画圆功能）；接着 onBind 方法绑定线程。在 Runnable 的 run 方法中，调用主程序 MainActivity 的 updateGUI 方法刷新界面。

DrawCircle.java 代码如下：

```java
package com.easybooks.android.circlebindservice;

import android.app.Service;
…

public class DrawCircle extends Service {
    private Thread drawThread;                              //声明线程对象
    private View circle;                                    //声明 MyCircle 对象
    private int radius;                                     //圆半径
    private static final String TAG = "绘圆服务（DrawCircle）"; //标识该服务

    private Runnable drawWork = new Runnable() {
        @Override
        public void run() {
            try {
                while (!drawThread.isInterrupted()) {
                    radius = (int) (Math.random() * 400);
                    MainActivity.updateGUI(circle, radius);
                    drawThread.sleep(1000);
                }
            } catch (InterruptedException e) {
                e.printStackTrace();
            }
        }
    };

    @Override
    public void onCreate() {
        super.onCreate();
        drawThread = new Thread(null, drawWork, "DrawCircle");
        circle = new MyCircle(this);
        Log.i(TAG, "onCreate");
    }

    public class CalcuCircletor extends Binder {            //(1)
```

```java
            DrawCircle getService() {
                return DrawCircle.this;
            }
        }

        @Override
        public IBinder onBind(Intent intent) {                    //(2)
            if (!drawThread.isAlive()) drawThread.start();
            Log.i(TAG, "onBind");
            return new CalcuCircletor();
        }

        @Override
        public boolean onUnbind(Intent intent) {                  //(3)
            drawThread.interrupt();
            Log.i(TAG, "onUnbind");
            return super.onUnbind(intent);
        }

        public class MyCircle extends View {
            public MyCircle(Context context) {
                super(context);
            }

            @Override
            protected void onDraw(Canvas canvas) {
                super.onDraw(canvas);
                Paint paint = new Paint();
                canvas.drawCircle(400, 400, radius, paint);
            }
        }

        /**
         * 公开提供给服务调用者使用的方法                          //(4)
         */
        public float getCircu() {
            return (float) (2 * Math.PI * radius);                //计算圆周长
        }

        public float getArea() {
            return (float) (Math.PI * radius * radius);           //计算圆面积
        }
}
```

其中：

（1）CalcuCircletor 继承自 Binder，在绑定方式使用的服务中必须实现这个类，它的作用在于以对象形式返回服务本身，以供主程序调用服务中对外公开的方法。

（2）onBind 方法不仅以绑定方式开启服务，还返回给主程序一个 CalcuCircletor 对象，主程序就是通过它获取到服务对象的。

（3）onUnbind 解除绑定，即终止服务线程。

（4）主程序就是通过这些方法与绑定服务交互的。这里对外提供 getCircu 和 getArea 方法分别实现计算圆周长和面积的功能。

3. 主程序绑定

在主程序的 updateGUI 方法中通过 Handler（句柄）对象的 post 方法提交界面刷新机制，而界面刷新的具体操作在 Runnable 接口 refreshWork 的 run 方法中执行。与上例不同的是，在 onCreate 中要创建 ServiceConnection 对象，并重写其 onServiceConnected 方法获取服务实例，这样就可以在点击"计算"按钮执行事件方法 getCircleParam 中调用服务所公开的计算圆周长和面积的 getCircu 和 getArea 方法。

MainActivity.java 代码如下：

```java
package com.easybooks.android.circlebindservice;

import android.content.ComponentName;
…

public class MainActivity extends AppCompatActivity {
    private static FrameLayout circleFrame;
    private static TextView textRadius;
    private static TextView textCircleParam;
    private static Handler handler = new Handler();         //界面刷新句柄
    private static View mycircle;                           //圆视图对象
    private static int myradius;                            //圆半径
    private DrawCircle drawCircle;                          //声明服务对象
    private ServiceConnection myconn;

    @Override
    protected void onCreate(Bundle savedInstanceState) {
        super.onCreate(savedInstanceState);
        setContentView(R.layout.activity_main);
        findViews();
        Toolbar toolbar = (Toolbar) findViewById(R.id.toolbar);
        …
        myconn = new ServiceConnection() {                  //(1)
            @Override
            public void onServiceConnected(ComponentName name, IBinder service) {
                drawCircle = ((DrawCircle.CalcuCircletor) service).getService();    //一旦绑定成功就获取服务实例
                Log.i("ServiceConnection", "已获取所绑定的服务实例，可进行数据交互");
            }

            @Override
            public void onServiceDisconnected(ComponentName name) {
                drawCircle = null;
            }
        };
    }

    private void findViews() {
        circleFrame = (FrameLayout) findViewById(R.id.circleFrame);
        textRadius = (TextView) findViewById(R.id.textRadius);
```

```java
        textCircleParam = (TextView) findViewById(R.id.textCircleParam);
    }

    public static void updateGUI(View c, int r) {
        mycircle = c;
        myradius = r;
        handler.post(refreshWork);
    }

    private static Runnable refreshWork = new Runnable() {
        @Override
        public void run() {
            //刷新界面
            circleFrame.removeAllViews();
            circleFrame.addView(mycircle);
            textRadius.setText("半径: " + String.valueOf(myradius));
        }
    };

    //绑定服务
    public void bind(View view) {
        Intent intent = new Intent(this, DrawCircle.class);
        bindService(intent, myconn, BIND_AUTO_CREATE);
    }

    //解绑服务
    public void unbind(View view) {
        unbindService(myconn);
        drawCircle = null;
        Log.i("MainActivity", "已销毁所解绑的服务实例,并断开数据连接");
    }

    //获取当前圆周长和面积数值显示在界面下方
    public void getCircleParam(View view) {
        if (drawCircle != null) {                          //(2)
            float circu = drawCircle.getCircu();
            float area = drawCircle.getArea();
            textCircleParam.setText("圆周长: " + String.format("%.2f", circu) + "\n" + "面积: " + String.format("%.4f", area));
        }
        else
            textCircleParam.setText("圆周长: 面积: ");
    }
    …
}
```

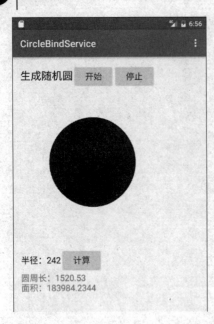

图 7.7 绑定绘圆服务的运行效果

其中：

（1）创建 ServiceConnection 对象的主要作用就是获取所绑定的服务实例，服务实例通过 DrawCircle.CalcuCircletor 的 getService 方法获取。ServiceConnection 对象还有一个 onServiceDisconnected 方法，在其中销毁已获取的服务实例。

（2）因为之前获取了服务实例，只要服务实例不为空，主程序就可以直接调用服务中公开的方法。

4. 服务配置

在 AndroidManifest.xml 中配置服务 DrawCircle，方法同前，此处略。

5. 运行效果

单击"开始"，服务线程开始以随机半径绘圆，单击"计算"，主程序从绑定的服务实例中获得当前圆的周长和面积值，并显示在屏幕下方。如图 7.7 所示。

读者同样也可由开发环境底部日志窗口的输出，观察到绑定服务的整个生命周期过程，如图 7.8 所示。

```
12-08 06:56:19.947 6380-6380/com.easybooks.android.circlebindservice I/绘图服务（DrawCircle）: onCreate
12-08 06:56:19.947 6380-6380/com.easybooks.android.circlebindservice I/绘图服务（DrawCircle）: onBind
12-08 06:56:19.953 6380-6380/com.easybooks.android.circlebindservice I/ServiceConnection: 已获取所绑定的服务实例,可进行数据交互
12-08 07:02:52.268 6380-6380/com.easybooks.android.circlebindservice I/MainActivity: 已销毁所解绑的服务实例,并断开数据连接
12-08 07:02:52.269 6380-6521/com.easybooks.android.circlebindservice W/System.err: java.lang.InterruptedException
12-08 07:02:52.269 6380-6521/com.easybooks.android.circlebindservice W/System.err:     at java.lang.Thread.sleep(Native Method)
12-08 07:02:52.269 6380-6521/com.easybooks.android.circlebindservice W/System.err:     at java.lang.Thread.sleep(Thread.java:1031)
12-08 07:02:52.269 6380-6521/com.easybooks.android.circlebindservice W/System.err:     at java.lang.Thread.sleep(Thread.java:985)
12-08 07:02:52.269 6380-6521/com.easybooks.android.circlebindservice W/System.err:     at com.easybooks.android.circlebindservice.DrawCircle$1.run(DrawCircle.java:29)
12-08 07:02:52.269 6380-6521/com.easybooks.android.circlebindservice W/System.err:     at java.lang.Thread.run(Thread.java:818)
12-08 07:02:52.269 6380-6380/com.easybooks.android.circlebindservice I/绘图服务（DrawCircle）: onUnbind
```

图 7.8 绑定服务的生命周期

7.2.4 多 Service 交互及生命周期

当我们第一次启动 Service 时，先后调用了 onCreate()和 onStart()这两个方法。当停止 Service 时，则执行 onDestroy()方法。如果 Service 已经启动了，当我们再次启动 Service 时，不会再执行 onCreate()方法，而是直接执行 onStart()方法。它可以通过 Service.stopSelf() 方法或者 Service.stopSelfResult()方法来停止自己。只要调用一次 stopService()方法便可以停止服务，无论调用了多少次启动服务方法。

（1）被启动服务的生命周期

如果一个 Service 被某个 Activity 调用 Context.startService 方法启动，那么不管是否有 Activity 使用 bindService 绑定或 unbindService 解除绑定到该 Service，该 Service 都在后台运行。如果一个 Service 被 startService 方法多次启动，那么 onCreate 方法只会调用一次，onStart 每次都会调用，并且系统只会创建 Service 的一个实例。该 Service 将会一直在后台运行，而不管对应程序的 Activity 是否在运行，直到被调用 stopService，或自身的 stopSelf 方法。当然如果系统资源不足，Android 系统也可能结束服务。

（2）被绑定服务的生命周期

如果一个 Service 被某个 Activity 调用 Context.bindService 方法绑定启动，不管调用 bindService 几

次，onCreate 方法都只会调用一次，同时 onStart 方法始终不会被调用。当连接建立之后，Service 将会一直运行，除非调用 Context.unbindService 断开连接或者之前调用 bindService 的 Context 不存在了，系统将会自动停止 Service，对应 onDestroy 将被调用。

（3）被启动又被绑定的服务的生命周期

如果一个 Service 又被启动又被绑定，则该 Service 将会一直在后台运行。并且不管如何调用，onCreate 始终只会调用一次，对应 startService 调用多少次，Service 的 onStart 便会调用多少次。调用 unbindService 将不会停止 Service，而必须调用 stopService 或 Service 的 stopSelf 来停止服务。

（4）当服务被停止时清除服务

当一个 Service 被终止（可能是调用 stopService 或者调用 stopSelf 或者不再有绑定的连接）时，onDestroy 方法将会被调用，在这里可以做一些清除工作，例如停止在 Service 中创建并运行的线程。

需要注意：

（1）调用 bindService 绑定到 Service 时，应保证在某处调用 unbindService 解除绑定。

（2）使用 startService 启动服务之后，不管是否使用 bindService，都要使用 stopService 停止服务。

（3）同时使用 startService 与 bindService 时，不管调用顺序如何，需要同时调用 unbindService 与 stopService 才能终止 Service。

（4）新 SDK 版本 onStart 方法已经变为 onStartCommand，但 onStart 仍然有效。

【例7.3】创建包含两个服务 PowerMonitor（电源监控）和 App（应用），主程序 MainActivity 初始默认以启动方式运行 PowerMonitor 服务，该服务在整个项目运行期间一直在后台，用户可单击屏幕上的"开启应用"按钮以启动方式运行 App 服务。App 启动后自动与 PowerMonitor 服务绑定，通过绑定来使用 PowerMonitor 所管理的电源。App 服务在运行期间也会定时查看 PowerMonitor 剩余的电量，若发现电量耗尽，就自行与 PowerMonitor 解除绑定并退出。用户可随时单击"关闭应用"来停止 App 服务。

1. 设计页面

创建 Android 工程 PowerMonitorNormal，其上的"开启应用"和"关闭应用"按钮是用来控制 App 服务的（见图7.9）。

content_main.xml 文件代码如下：

图7.9 设计视图

```
<?xml version="1.0" encoding="utf-8"?>
<RelativeLayout xmlns:android="http://schemas.android.com/apk/res/android"
    …>

    <TextView
        android:layout_width="wrap_content"
        android:layout_height="wrap_content"
        android:text="电量："
        android:id="@+id/textView"
        android:textColor="@android:color/black"
        android:textSize="@dimen/abc_text_size_large_material" />

    <ProgressBar
        style="?android:attr/progressBarStyleHorizontal"
```

```xml
        android:layout_width="match_parent"
        android:layout_height="100dp"
        android:id="@+id/progressBarPower"
        android:layout_alignBottom="@+id/textView"
        android:layout_toRightOf="@+id/textView"
        android:layout_toEndOf="@+id/textView"
        android:max="100"
        android:background="@android:color/holo_green_light" />

    <Button
        android:layout_width="match_parent"
        android:layout_height="wrap_content"
        android:text="开    启    应    用"
        android:id="@+id/buttonStartApp"
        android:layout_below="@+id/textView"
        android:layout_alignParentLeft="true"
        android:layout_alignParentStart="true"
        android:textSize="@dimen/abc_text_size_medium_material"
        android:layout_marginTop="20dp"
        android:onClick="start" />

    <Button
        android:layout_width="match_parent"
        android:layout_height="wrap_content"
        android:text="关    闭    应    用"
        android:id="@+id/buttonStopApp"
        android:layout_below="@+id/buttonStartApp"
        android:layout_alignParentLeft="true"
        android:layout_alignParentStart="true"
        android:textSize="@dimen/abc_text_size_medium_material"
        android:onClick="stop" />

</RelativeLayout>
```

2. 创建服务

（1）创建服务类 PowerMonitor

在它的 onCreate 中创建该服务线程并赋值初始电量为 100，线程启动后通过 Runnable 接口的 run 方法每 100 毫秒更新一次界面进度条的电量值。由于 App 服务需要与 PowerMonitor 绑定，故该服务中同时也实现了 onBind、onUnbind，并对外提供了一些公开使用的方法：checkPower 和 usePower。

PowerMonitor.java 代码如下：

```java
package com.easybooks.android.powermonitornormal;

import android.app.Service;
…
public class PowerMonitor extends Service {
    private Thread monitorThread;                                          //声明监控服务线程
    private int power;                                                     //剩余电量值
    private static final String TAG = "电源监控服务（PowerMonitor）";      //服务标识

    private Runnable monitorWork = new Runnable() {                        //①
```

```java
    @Override
    public void run() {
        try {
            while (!monitorThread.isInterrupted()) {
                MainActivity.updateGUI(power);
                monitorThread.sleep(100);
            }
        } catch (InterruptedException e) {
            e.printStackTrace();
        }
    }
};

@Override
public void onCreate() {
    super.onCreate();
    monitorThread = new Thread(null, monitorWork, "PowerMonitor");
    power = 100;
    Log.i(TAG, "onCreate");
}

@Override
public int onStartCommand(Intent intent, int flags, int startId) {
    super.onStartCommand(intent, flags, startId);
    if (!monitorThread.isAlive()) monitorThread.start();
    Log.i(TAG, "onStartCommand");
    return START_STICKY;
}

@Override
public void onDestroy() {
    super.onDestroy();
    monitorThread.interrupt();
    Log.i(TAG, "onDestroy");
}

public class Plug extends Binder {                              //②
    PowerMonitor getService() {
        return PowerMonitor.this;                               //模拟一个插头插到插座上获取电能
    }
}

@Override
public IBinder onBind(Intent intent) {
    if (!monitorThread.isAlive()) monitorThread.start();
    Log.i(TAG, "onBind");
    return new Plug();
}

@Override
public boolean onUnbind(Intent intent) {
```

```java
            Log.i(TAG, "onUnbind");
            return super.onUnbind(intent);
        }

        /**
         * 公开提供给服务调用者(应用服务)使用的方法                    //③
         */
        public int checkPower() {
            return power;//获取并查看电量值
        }

        public void usePower(int quantity) {
            power -= quantity;//修改电量值（模拟应用耗电）
            if (power < 0) power = 0;
        }
    }
```

其中：

① 电源监控服务与之前的服务一样，也是通过 Runnable 接口机制更新界面的，每隔 100 毫秒调用一次主程序界面的 updateGUI 方法。

② 定义类 Plug 继承自 Binder，其作用与上例的 CalcuCircletor 一样，也是为了让其他程序或服务能通过它来获取本服务的实例。

③ 这里定义了两个方法：checkPower 和 usePower，旨在公开让应用 App 服务调用，模拟查看剩余电量和耗电应用。

（2）创建服务类 App

该服务模拟其他耗电的手机应用，在 onCreate 中创建 ServiceConnection 对象获取电源监控服务实例，在 onStartCommand 中与电源监控服务绑定，App 服务运行后在 Runnable 接口中每秒调用一次监控服务的 usePower 耗电量 5，同时调用 checkPower 检查剩余电量值，一旦发现电量值小于 0，App 会自己调用 unbindService 与监控服务解除绑定，这样就不再消耗监控服务中的电了。

App.java 代码如下：

```java
package com.easybooks.android.powermonitornormal;

import android.app.Service;
…

public class App extends Service {
    private Thread appThread;                                   //声明应用服务线程
    private static final String TAG = "应用服务（App）";          //服务标识
    private PowerMonitor powerMonitor;                          //电源监控服务对象
    private ServiceConnection myconn;

    @Override
    public IBinder onBind(Intent intent) {
        return null;
    }

    private Runnable appWork = new Runnable() {                 //①
        @Override
        public void run() {
```

```java
            try {
                while (!appThread.isInterrupted()) {
                    if (powerMonitor != null) {
                        if (powerMonitor.checkPower() > 0) powerMonitor.usePower(5);//每秒耗电量为5
                        else {
                            unbindService(myconn);
                            powerMonitor = null;
                            Log.i(TAG, "应用监测到手机电已耗尽，自行退出");
                        }
                    }
                    appThread.sleep(1000);
                }
            } catch (InterruptedException e) {
                if (powerMonitor != null) {
                    unbindService(myconn);
                    powerMonitor = null;
                }
                Log.i(TAG, "应用被关闭");
                e.printStackTrace();
            }
        }
    };

    @Override
    public void onCreate() {
        super.onCreate();
        appThread = new Thread(null, appWork, "App");
        myconn = new ServiceConnection() {
            @Override
            public void onServiceConnected(ComponentName name, IBinder service) {
                powerMonitor = ((PowerMonitor.Plug) service).getService();//一旦绑定成功就获取服务实例
                Log.i("ServiceConnection", "应用已获取电源监控服务实例，可使用手机电源");
            }

            @Override
            public void onServiceDisconnected(ComponentName name) {
                powerMonitor = null;
            }
        };
        Log.i(TAG, "onCreate");
    }

    @Override
    public int onStartCommand(Intent intent, int flags, int startId) {
        super.onStartCommand(intent, flags, startId);
        Intent i = new Intent(this, PowerMonitor.class);
        bindService(i, myconn, BIND_AUTO_CREATE);                //②
        if (!appThread.isAlive()) appThread.start();
        Log.i(TAG, "onStartCommand");
        return START_STICKY;
    }
}
```

```
    @Override
    public void onDestroy() {
        super.onDestroy();
        appThread.interrupt();
        Log.i(TAG, "onDestroy");
    }
}
```

其中:

① App 应用服务也是通过 Runnable 接口机制运行，不断耗电和检查剩余电量，直到发现电量耗尽后与电源监控服务解绑。

② App 服务就是在这里绑定电源监控服务 PowerMonitor 的。

3. 主程序启动

主程序在 findViews 中启动 PowerMonitor 服务，界面进度条动态更新的机制与之前一样也是采用句柄提交 Runnable 接口的方式。通过界面上的"开启应用"按钮执行 start 方法启动 App 服务，"关闭应用"方法停止 App 服务。

MainActivity.java 代码如下:

```java
package com.easybooks.android.powermonitornormal;

import android.content.Intent;
...

public class MainActivity extends AppCompatActivity {
    private static ProgressBar progressBarPower;                //界面电量进度条
    private static Button buttonStartApp;
    private static Button buttonStopApp;
    private static Handler handler = new Handler();             //界面刷新句柄
    private static int mypower;                                 //电量值

    @Override
    protected void onCreate(Bundle savedInstanceState) {
        super.onCreate(savedInstanceState);
        setContentView(R.layout.activity_main);
        findViews();
        Toolbar toolbar = (Toolbar) findViewById(R.id.toolbar);
        ...
    }

    private void findViews() {
        progressBarPower = (ProgressBar) findViewById(R.id.progressBarPower);
        buttonStartApp = (Button) findViewById(R.id.buttonStartApp);
        buttonStopApp = (Button) findViewById(R.id.buttonStopApp);
        Intent intent = new Intent(this, PowerMonitor.class);
        startService(intent);                                   //初始启动 PowerMonitor 服务
    }

    public static void updateGUI(int p) {
        mypower = p;
```

```
            handler.post(refreshWork);
        }

        private static Runnable refreshWork = new Runnable() {
            @Override
            public void run() {
                //刷新界面
                progressBarPower.setProgress(mypower);
            }
        };

        //开启应用服务
        public void start(View view) {
            Intent intent = new Intent(this, App.class);
            startService(intent);
        }

        //停止应用服务
        public void stop(View view) {
            Intent intent = new Intent(this, App.class);
            stopService(intent);
        }
        …
}
```

4. 服务配置

在 AndroidManifest.xml 中，将开发好的两个服务（PowerMonitor 和 App）配置其中（加黑处）。

```xml
<?xml version="1.0" encoding="utf-8"?>
<manifest xmlns:android="http://schemas.android.com/apk/res/android"
    package="com.easybooks.android.powermonitornormal">

    <application
        …
        <activity
            android:name=".MainActivity">
            …
        </activity>
        <service android:name=".PowerMonitor"/>
        <service android:name=".App"/>
    </application>

</manifest>
```

5. 运行效果

运行程序后，单击"开启应用"，上方显示电量（红色进度条）开始逐步减少。单击"关闭应用"，电量停止减少。再次单击"开启应用"，电量值又继续减少，直到红色条减退消失（表示电量耗尽），应用自行退出。此时若再单击"开启应用"，应用刚一启动就会立马退出，反复启动皆是如此（因为已经没电了）。如图 7.10 所示。

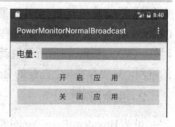

图7.10　多服务交互运行效果

整个过程可从下面两个服务的生命周期日志中很清楚地看出来，如图 7.11 所示。

```
12-08 08:41:24.339 17810-17810/com.easybooks.android.powermonitornormalbroadcast I/电源监控服务 (PowerMonitor): onCreate
12-08 08:41:24.339 17810-17810/com.easybooks.android.powermonitornormalbroadcast I/电源监控服务 (PowerMonitor): onStartCommand
                                                          [ 12-08 08:41:24.369 17810:17854 D/         ]
                                                          HostConnection::get() New Host Connection established 0xaa3fafa0, tid 17854
12-08 08:41:24.372 17810-17854/com.easybooks.android.powermonitornormalbroadcast I/OpenGLRenderer: Initialized EGL, version 1.4
12-08 08:41:24.395 17810-17854/com.easybooks.android.powermonitornormalbroadcast W/EGL_emulation: eglSurfaceAttrib not implemented
12-08 08:41:24.395 17810-17854/com.easybooks.android.powermonitornormalbroadcast W/OpenGLRenderer: Failed to set EGL_SWAP_BEHAVIOR on surface 0xaa00fa80, error=EGL_SUCCESS
12-08 08:41:28.455 17810-17810/com.easybooks.android.powermonitornormalbroadcast I/应用服务 (App): onCreate
12-08 08:41:28.456 17810-17810/com.easybooks.android.powermonitornormalbroadcast I/应用服务 (App): onStartCommand
12-08 08:41:28.456 17810-17810/com.easybooks.android.powermonitornormalbroadcast I/电源监控服务 (PowerMonitor): onBind
12-08 08:41:28.457 17810-17810/com.easybooks.android.powermonitornormalbroadcast I/ServiceConnection: 应用已获取电源监控服务实例，可使用手机电源
12-08 08:41:33.137 17810-17810/com.easybooks.android.powermonitornormalbroadcast I/应用服务 (App): onDestroy
12-08 08:41:33.140 17810-17810/com.easybooks.android.powermonitornormalbroadcast I/电源监控服务 (PowerMonitor): onUnbind
12-08 08:41:33.140 17810-17918/com.easybooks.android.powermonitornormalbroadcast I/应用服务 (App): 应用被关闭
12-08 08:41:38.419 17810-17810/com.easybooks.android.powermonitornormalbroadcast I/应用服务 (App): onCreate
12-08 08:41:38.420 17810-17810/com.easybooks.android.powermonitornormalbroadcast I/应用服务 (App): onStartCommand
12-08 08:41:38.420 17810-17810/com.easybooks.android.powermonitornormalbroadcast I/ServiceConnection: 应用已获取电源监控服务实例，可使用手机电源
12-08 08:41:53.439 17810-18066/com.easybooks.android.powermonitornormalbroadcast I/应用服务 (App): 应用监测到手机电已耗尽，自行退出
12-08 08:41:57.302 17810-17810/com.easybooks.android.powermonitornormalbroadcast I/应用服务 (App): onStartCommand
12-08 08:41:57.318 17810-17810/com.easybooks.android.powermonitornormalbroadcast I/ServiceConnection: 应用已获取电源监控服务实例，可使用手机电源
12-08 08:41:57.444 17810-18066/com.easybooks.android.powermonitornormalbroadcast I/应用服务 (App): 应用监测到手机电已耗尽，自行退出
12-08 08:42:03.452 17810-17810/com.easybooks.android.powermonitornormalbroadcast I/应用服务 (App): onStartCommand
12-08 08:42:03.452 17810-17810/com.easybooks.android.powermonitornormalbroadcast I/ServiceConnection: 应用已获取电源监控服务实例，可使用手机电源
12-08 08:42:04.453 17810-18066/com.easybooks.android.powermonitornormalbroadcast I/应用服务 (App): 应用监测到手机电已耗尽，自行退出
```

图 7.11　各个服务的生命周期日志

7.3　广播（BroadcastReceiver）

7.3.1　BroadcastReceiver 概述

在 Android 系统中，广播体现在方方面面，例如当开机完成后系统会产生一条广播，接收到这条广播就能实现开机启动服务的功能。当网络状态改变时系统会产生一条广播，接收到这条广播就能及时地做出提示和保存数据等操作。当电池电量改变时，系统会产生一条广播，接收到这条广播就能在电量低时告知用户及时保存进度。

Android 中的广播机制非常出色，很多事情原本需要开发者亲自操作的，现在只需等待广播告知自己就可以了，大大减少了开发的工作量和开发周期。

1. 基本概念

BroadcastReceiver（广播接收者）用于异步接收广播，广播的发送是通过调用 Context.sendBroadcast()、Context.sendOrderedBroadcast()、Context.sendStickyBroadcast()来实现的。通常一个广播可以被订阅了此 Intent 的多个广播接收者所接收。广播接收器通过调用 BroadcastReceiver()方法接收广播，对广播的通知做出反应。广播接收器没有用户界面，但是它可以为它们接收到信息启动一个 Activity 或者使用 NotificationManager 来通知用户。

BroadcastReceiver 是一个系统全局的监听器，用于监听系统全局的 Broadcast 消息，所以它可以很方便地进行系统组件之间的通信。BroadcastReceiver 虽然是一个监听器，但是它和之前用到的 OnXxxListener 监听器不同，那些只是程序级别的监听器，运行在指定程序的所在进程中，当程序退出的时候，OnXxxListener 监听器也就随之关闭了。但是 BroadcastReceiver 属于系统级的监听器，它拥有自己的进程，只要存在与之匹配的 Broadcast 被以 Intent 的形式发送出来，BroadcastReceiver 就会被激活。

2. 生命周期

虽然 BroadcastReceiver 同属 Android 的四大组件，但它也有自己独立的生命周期，但是和 Activity

与 Service 又不同。当在系统注册一个 BroadcastReceiver 之后，每次系统以一个 Intent 的形式发布 Broadcast 的时候，系统都会创建与之对应的 BroadcastReceiver 广播接收者实例，并自动触发它的 onReceive()方法。当 onReceive()方法被执行完成之后，BroadcastReceiver 实例就会被销毁。虽然它独自享用一个单独的进程，但也不是没有限制的，如果 BroadcastReceiver.onReceive()方法不能在几秒内执行完成，Android 系统就会认为该 BroadcastReceiver 对象无响应，然后弹出"Application No Response"对话框，所以不要在 BroadcastReceiver.onReceive()方法内执行一些耗时的操作。

BroadcastReceiver 的使用基本流程如下：注册 receiver（静态或动态）→调用 sendBroadcast()方法发送广播→调用 onReceive(Context context，Intent intent)方法处理广播→调用 startService()方法启动服务→调用 stopService()方法关闭服务。

一个 BroadcastReceiver 对象只有在被调用 onReceive(Context，Intent)方法时才有效，当从该方法返回后，该对象就无效了，其生命周期结束。

从这个特征可以看出，在所调用的 onReceive(Context，Intent)方法里，不能有过于耗时的操作，不能使用线程来执行。对于耗时的操作，应该在 startService 中来完成。因为当得到其他异步操作所返回的结果时，BroadcastReceiver 可能已经无效了。

如果需要根据广播内容完成一些耗时的操作，一般考虑通过 Intent 启动一个 Service 来完成该操作，而不应该在 BroadcastReceiver 中开启一个新线程完成耗时的操作，因为 BroadcastReceiver 本身的生命周期很短，可能出现的情况是子线程还没有结束，BroadcastReceiver 就已经退出的情况，而如果 BroadcastReceiver 所在的进程结束了，该线程就会被标记为一个空线程。根据 Android 的内存管理策略，在系统内存紧张的时候，会按照优先级，结束优先级低的线程，而空线程是优先级最低的，这样就可能导致 BroadcastReceiver 启动的子线程不能执行完成。

3. 注册方式

BroadcastReceiver 本质上是一个监听器，所以使用 BroadcastReceiver 的方法也非常简单，只需要继承 BroadcastReceiver，在其中重写 onReceive(Context context,Intent intent)即可。一旦实现了 BroadcastReceiver，并部署到系统中后，就可以在系统的任何位置，通过 sendBroadcast、sendOrderedBroadcast 方法发送 Broadcast 给这个 BroadcastReceiver。

但是仅仅继承 BroadcastReceiver 和实现 onReceive()方法是不够的，同为 Android 系统组件，它也必须在 Android 系统中注册。注册一个 BroadcastReceiver 有以下两种方式。

（1）静态注册

在 AndroidManifest.xml 文件的<application/>节点中使用标签<receiver>进行注册，并在标签内用<intent- filter>标签设置过滤器。例如：

```
<receiver android:name="myReceiver">
    <intent-filter>
        <action android:name="android.intent.action.myBroadcast"/>
        <category android:name="android.intent.category.DEFAULT"/>
    </intent-filter>
</receiver>
```

其中：

在<receiver/>节点中用 android:name 属性中指定注册的 BroadcastReceiver 对象，通过<Intent-filter>指定<action>和<category>，并可通过 android:priority 属性设置 BroadcastReceiver 的优先级，数值越大优先级越高。

此后，只要是 android.intent.action. myBroadcast 地址的广播，myReceiver 都能够接收到。

注意，这种方式的注册是常驻型的，也就是说当应用关闭后，如果有广播信息传来，MyReceiver 也会被系统调用而自动运行。

静态注册方式，由系统来管理 receiver，而且程序里的所有 receiver 可以在 XML 文件中一目了然。

（2）动态注册

动态注册需要在代码中动态地指定广播地址并注册，在代码中先定义并设置好一个 IntentFilter 对象，然后在需要注册的地方调用 Context.registerReceiver()方法，需要取消时就调用 Context.unregisterReceiver()方法。当用动态方式注册的 BroadcastReceiver 的 Context 对象被销毁时，BroadcastReceiver 也就自动取消注册了。

另外，如果在使用 sendBroadcast()方法时指定了接收权限，则只有在 AndroidManifest.xml 文件中用<uses- permission>标签声明了拥有此权限的 BroadcastReceiver 才会有可能接收到发送来的 Broadcast。同样，如果在注册 BroadcastReceiver 时指定了可接收的 Broadcast 的权限，则只有在 AndroidManifest.xml 文件中用<uses-permission>标签声明了，拥有此权限的 Context 对象所发送的 Broadcast 才能被这个 BroadcastReceiver 所接收。

例如：

```
myReceiver receiver = new myReceiver();
IntentFilter filter = new IntentFilter();
Filter.addAction("android.intent.action.myBroadcast");
registerReceiver(receiver,filter);
```

注意，registerReceiver 是 android.content.ContextWrapper 类中的方法，Activity 和 Service 都继承了 ContextWrapper 类，所以可以直接调用。在实际应用中，我们在 Activity 或 Service 中注册了一个 BroadcastReceiver，当这个 Activity 或 Service 被销毁时如果没有解除注册，系统会报一个异常，提示我们是否忘记解除注册了。所以，记得在特定的地方执行解除注册操作，解除注册的代码如下所示：

```
@override
Protected void onDestroy(){
    super.onDestroy();
    unregisterReceiver(receiver);
}
```

注意，动态注册方式不是常驻型的，广播会跟随程序的生命周期。

动态注册方式的注册代码隐藏在代码中，一般在 activity 的 onStart()里面进行注册，在 onStop()里面进行注销。在退出程序前要记得调用 Context.unregisterReceiver()方法。如果在 Activity.onResume()里面注册了，就必须在 Activity.onPause()注销。

虽然 Android 系统提供了两种方式注册 BroadcastReceiver，但是一般在实际开发中，还是会在清单文件 AndroidManifest.xml 进行注册。

4. 广播类型

根据 Broadcast 的传播方式，在系统中有两种类型的 Broadcast。

（1）普通广播（Normal Broadcast）

普通广播对于多个接收者来说是完全异步的，通常每个接收者都无须等待即可接收到广播，接收者相互之间不会有影响。对于这种广播，接收者无法终止广播，即无法阻止其他接收者的接收动作。当一个 Broadcast 被发出之后，所有与之匹配的 BroadcastReceiver 都同时接收到 Broadcast。这种方式的优点是传递效率比较高，其缺点是一个 BroadcastReceiver 不能影响其他响应这个 Broadcast 的 BroadcastReceiver。

发送普通广播的方法：sendBroadcast();

注意，sendBroadcast 也是 android.content.ContextWrapper 类中的方法，它可以将一个指定地址和参数信息的 Intent 对象以广播的形式发送出去。

（2）有序广播（Ordered Broadcast）

有序广播是同步执行的，也就是说有序广播的接收器将会按照预先声明的优先级依次接受 Broadcast，它是链式结构，每次只发送到优先级较高的接收者那里，优先级越高越先被执行。因为是顺序执行，所有优先级高的接收器可以把执行结果传播到优先级低的接收者那里，优先级高的接收者有能力终止这个广播，可以通过 abortBroadcast()方法终止 Broadcast 的传播。一旦 Broadcast 的传播被终止，优先级低于它的接收器就不会再接收到这条 Broadcast 了。

发送有序广播的方法：sendOrderedBroadcast()。

7.3.2 普通广播应用

【例 7.4】在上面多服务交互实例的基础上，应用普通广播在手机电量快耗尽（20%以下）时给予提示，告知用户电量不足。

1. 创建电源监控服务和应用服务

创建工程 PowerMonitorNormalBroadcast，其他页面和类代码与 PowerMonitorNormal 相同。

2. 创建广播接收器

在工程中添加一个广播接收器类 PowerBroadcast，PowerBroadcast.java 文件代码如下：

```java
package com.easybooks.android.powermonitornormalbroadcast;

import android.content.BroadcastReceiver;
import android.content.Context;
import android.content.Intent;
import android.widget.Toast;

public class PowerBroadcast extends BroadcastReceiver {
    @Override
    public void onReceive(Context context, Intent intent) {
        Toast.makeText(context, intent.getStringExtra("msg"), Toast.LENGTH_SHORT).show();
    }
}
```

该类继承自 BroadcastReceiver，实现了其 onReceive 方法，其中仅有一条语句用于显示广播消息，消息经由 Intent 类型的参数传递。

3. 配置广播接收器

在 AndroidManifest.xml 中，为该广播接收器添加配置（加黑处）如下：

```xml
<?xml version="1.0" encoding="utf-8"?>
<manifest xmlns:android="http://schemas.android.com/apk/res/android"
    package="com.easybooks.android.powermonitornormalbroadcast">

    <application
        …
        <activity
            android:name=".MainActivity"
            …
        </activity>
```

```xml
            <service android:name=".PowerMonitor"/>
            <service android:name=".App"/>
            <receiver android:name=".PowerBroadcast">
                <intent-filter>
                    <action android:name="com.easybooks.android.powermonitornormalbroadcast"/>
                </intent-filter>
            </receiver>
    </application>

</manifest>
```

4. 发送广播

在服务类 PowerMonitor 中添加发送广播的代码,位于 PowerMonitor.java 中(加黑处),如下:

```java
package com.easybooks.android.powermonitornormalbroadcast;

import android.app.Service;
…
public class PowerMonitor extends Service {
    private Thread monitorThread;
    private int power;
    private boolean isSend = false;                   //是否已发过广播
    private static final String TAG = "电源监控服务(PowerMonitor)";

    private Runnable monitorWork = new Runnable() {
        @Override
        public void run() {
            try {
                while (!monitorThread.isInterrupted()) {
                    MainActivity.updateGUI(power);
                    monitorThread.sleep(100);
                    //电量不足 20%时以广播方式告知主程序
                    if (power < 20 && !isSend) {
                        Intent intent = new Intent();
                        intent.setAction("com.easybooks.android.powermonitornormalbroadcast");
                                                                    //(1)
                        intent.putExtra("msg","电池电量不足 20%! ");
                        sendBroadcast(intent);                      //(2)
                        isSend = true;
                    }
                }
            } catch (InterruptedException e) {
                e.printStackTrace();
            }
        }
    };
    …
}
```

其中：

（1）"com.easybooks.android.powermonitornormalbroadcast"为广播消息的接收地址，该地址必须与之前 AndroidManifest.xml 中配置的"<action android:name="com.easybooks.android. powermonitornormal broadcast"/>"中的 name 值相一致。

（2）sendBroadcast 方法发送广播消息。

5. 运行效果

运行程序，单击"开启应用"，当电量条减少到 20%时，系统弹出消息提示电量不足。如图 7.12 所示。

图 7.12 普通广播的运行效果

7.3.3 有序广播应用

【例 7.5】虽然系统存在两种类型的 Broadcast，但是一般系统发送出来的 Broadcast 均是有序广播，所以可以通过优先级的控制，在系统内置的程序响应前，对 Broadcast 提前进行响应。这就是短信拦截器、电话拦截器等软件的工作原理。

1. 设计页面

创建 Android 工程 SmsFilterOrderedBroadcast。

content_main.xml 文件代码如下：

```xml
<?xml version="1.0" encoding="utf-8"?>
<RelativeLayout xmlns:android="http://schemas.android.com/apk/res/android"
        …>

    <TextView
        android:layout_width="wrap_content"
        android:layout_height="wrap_content"
        android:textSize="@dimen/abc_text_size_large_material"
        android:textColor="@android:color/black"
        android:id="@+id/textSmsMsg" />
</RelativeLayout>
```

该页面很简单，上面仅一个 TextView 用于显示短信消息。

2. 创建服务

创建服务类 SmsSender 用于发送短信，该类模拟的是一个基站，通过 Runnable 接口不断地产生和发送短信，在 onCreate 中创建服务线程，在 onStartCommand 中启动线程。

SmsSender.java 代码如下：

```java
package com.easybooks.android.smsfilterorderedbroadcast;

import android.app.Service;
…

public class SmsSender extends Service {
    private Thread sendThread;                         //声明短信发送线程
    private String[] telHead = {"138", "170"};         //短信号码头
    private String[] smsContent = {"周何骏先生，祝生日快乐！", "恭喜您中了 500 万大奖！"};
                                                       //短信内容
    private String[] smsMessages = new String[2];      //存储短信结构（号码+信息内容）
```

```java
    private static final String TAG = "短信发送服务（SmsSender）";   //服务标识

    @Override
    public IBinder onBind(Intent intent) {
        return null;
    }

    private Runnable sendWork = new Runnable() {
        @Override
        public void run() {
            try {
                while (!sendThread.isInterrupted()) {
                    for (int i = 0; i < 2; i++) {
                        int number = new Random().nextInt(89999999) + 10000000;  //(1)
                        smsMessages[0] = telHead[i] + String.valueOf(number);
                        smsMessages[1] = smsContent[i];                          //(2)
                        Intent intent = new Intent();
                        intent.setAction("com.easybooks.android.smsfilterorderedbroadcast");
                        intent.putExtra("tel", smsMessages[0]);
                        intent.putExtra("sms", smsMessages[1]);
                        sendOrderedBroadcast(intent, null);   //发送有序广播
                        sendThread.sleep(5000);
                    }
                }
            } catch (InterruptedException e) {
                e.printStackTrace();
            }
        }
    };

    @Override
    public void onCreate() {
        super.onCreate();
        sendThread = new Thread(null, sendWork, "SmsSender");
        Log.i(TAG, "onCreate");
    }

    @Override
    public int onStartCommand(Intent intent, int flags, int startId) {
        super.onStartCommand(intent, flags, startId);
        if (!sendThread.isAlive()) sendThread.start();
        Log.i(TAG, "onStartCommand");
        return START_STICKY;
    }

    @Override
    public void onDestroy() {
        super.onDestroy();
        sendThread.interrupt();
        Log.i(TAG, "onDestroy");
    }
}
```

其中：
（1）短信号码是随机产生的，加上 138 或 170 前缀，形成完整的模拟手机号。
（2）短信号码和信息内容分别作为一个字段，存储在一个包含两个元素的字符串数组 smsMessages 中，并进一步封装入 Intent 待发送。

3. 创建短信接收器

创建广播接收器类 SmsReceiver 用于接收短信，SmsReceiver.java 代码如下：

```
package com.easybooks.android.smsfilterorderedbroadcast;

import android.content.BroadcastReceiver;
import android.content.Context;
import android.content.Intent;

public class SmsReceiver extends BroadcastReceiver {
    @Override
    public void onReceive(Context context, Intent intent) {
        String smsMsg = "来自" + intent.getStringExtra("tel") + "：\n" + intent.getStringExtra("sms");
        MainActivity.updateGUI(smsMsg);
    }
}
```

在 onReceive 方法中调用主程序的 updateGUI 刷新界面，在主界面的 TextView 中显示短信。

4. 创建短信拦截器

创建短信拦截器类 SmsFilter 用于拦截 170 打头（现实生活中多为诈骗）的短信，通过 abortBroadcast 方法拦截短信。

SmsFilter.java 代码如下：

```
package com.easybooks.android.smsfilterorderedbroadcast;

import android.content.BroadcastReceiver;
import android.content.Context;
import android.content.Intent;
import android.widget.Toast;

public class SmsFilter extends BroadcastReceiver {
    @Override
    public void onReceive(Context context, Intent intent) {
        String telHead = intent.getStringExtra("tel").substring(0, 3);
        if (telHead.equals("170")) {
            abortBroadcast();           //中断广播，不会再向比它优先级低的接收器传播下去了
            Toast.makeText(context, "拦截到一个诈骗短信！\n 来自" + intent.getStringExtra("tel"), Toast.LENGTH_SHORT).show();
        }
    }
}
```

5. 主程序

主程序在 findViews 方法中启动短信发送服务 SmsSender，通过 updateGUI 刷新界面上显示的短信息，界面刷新同样采用的是 Runnable 机制。

MainActivity.java 代码如下：

```java
package com.easybooks.android.smsfilterorderedbroadcast;

import android.content.Intent;
…

public class MainActivity extends AppCompatActivity {
    private static TextView textSmsMsg;
    private static Handler handler = new Handler();        //界面刷新句柄
    private static String mySmsMsg;                        //短信内容

    @Override
    protected void onCreate(Bundle savedInstanceState) {
        super.onCreate(savedInstanceState);
        setContentView(R.layout.activity_main);
        findViews();
        Toolbar toolbar = (Toolbar) findViewById(R.id.toolbar);
    …
    }

    private void findViews() {
        textSmsMsg = (TextView) findViewById(R.id.textSmsMsg);
        Intent intent = new Intent(this, SmsSender.class);
        startService(intent);                              //启动短信发送服务
    }

    public static void updateGUI(String msg) {             //刷新短信
        mySmsMsg = msg;
        handler.post(refreshWork);
    }

    private static Runnable refreshWork = new Runnable() {
        @Override
        public void run() {
            //刷新界面
            textSmsMsg.setText(mySmsMsg);
        }
    };
    …
}
```

6. 接收与拦截配置

在 AndroidManifest.xml 中，短信拦截器（SmsFilter）的优先级必须配置高于短信接收器（SmsReceiver），才能发挥拦截作用，代码如下：

```xml
<?xml version="1.0" encoding="utf-8"?>
<manifest xmlns:android="http://schemas.android.com/apk/res/android"
    package="com.easybooks.android.smsfilterorderedbroadcast">

    <application
        …
```

```xml
<activity
    android:name=".MainActivity"
    …
</activity>
<service android:name=".SmsSender"/>
<!--priority 优先级：数字越高优先级越高-->
<receiver android:name=".SmsFilter">
    <intent-filter android:priority="10000">
        <action android:name="com.easybooks.android.smsfilterorderedbroadcast"/>
    </intent-filter>
</receiver>
<receiver android:name=".SmsReceiver">
    <intent-filter android:priority="10">
        <action android:name="com.easybooks.android.smsfilterorderedbroadcast"/>
    </intent-filter>
</receiver>
</application>
</manifest>
```

7. 运行效果

运行程序，当不配置拦截器时，主程序将同时收到 138 和 170 打头的短信，如图 7.13 所示。
当配置拦截器时，主程序将收到 138 的短信，收到了拦截 170 打头的短信广播，如图 7.14 所示。

图 7.13　未配置拦截　　　　　　　　　　图 7.14　使用拦截器屏蔽 170 短信

第8章 Android 数据存储与共享

Android 系统提供多种数据存储方法，包括易于使用的 SharedPreferences、经典的文件存储和轻量级的 SQLite 数据库。

不同的数据存储方法有着不同的适用领域和跨应用的数据共享方法。

8.1 SharedPreferences（共享优先）存储

8.1.1 SharedPreferences 概述

SharedPreferences 称为共享优先存储，是一种轻量级的数据保存方式，通过调用函数就可以实现对 NVP（名称/值对）的保存和读取。

SharedPreferences 不仅能够保存数据，还能够实现不同应用程序间的数据共享。它支持三种访问模式：私有（MODE_PRIVATE）、全局读（MODE_WORLD_READABLE）和全局写（MODE_WORLD_WRITEABLE）。如果定义为私有模式，仅创建的程序有权限对其进行读取或写入。如果定义为全局读模式，不仅创建程序可以对其进行读取或写入，其他应用程序也具有读取操作的权限，但没有写入操作的权限。如果定义为全局写模式，则所有程序都可以对其进行写入操作，但没有读取操作的权限。

（1）在使用 SharedPreferences 前，先定义 SharedPreferences 的访问模式。

将访问模式定义为私有模式：

public static int MODE=MODE_PRIVATE;

访问模式设定为既可以全局读，也可以全局写：

public static int MODE = Context.MODE_WORLD_READABLE+ Context.MODE_WORLDWRITEABLE;

（2）定义 SharedPreferences 的名称，这个名称也是它在 Android 文件系统中保存的文件名。一般声明为字符串常量，这样可以在代码中多次使用：

public static final String PREFERENCE NAME= "SaveSetting";

（3）使用时需要将访问模式和名称作为参数传递到 getSharedPreferences()函数，则可获取 SharedPreferences 实例。

SharedPreferences   sharedPreferences = getSharedPreferences (PREFERENCENAME, MODE);

（4）在获取实例后，可以通过 SharedPreferences.Editor 类进行修改，最后调用 commit()函数保存修改内容。

SharedPreferences 广泛支持各种基本数据类型，包括整型、布尔型、浮点型和长型等。

```
SharedPreferences.Editor editor = sharedPreferences.edit();
editor.putString("Name", "Tom");
editor.putInt("Age", 20);
editor.putFloat("Height", 1.81f);
editor.commit();
```

（5）如果需要从已经保存的 SharedPreferences 中读取数据,同样是调用 getSharedPreferences()函数，

并在函数第 1 个参数中指明需要访问的 SharedPreferences 名称，最后通过 get<Type>()函数获取保存在 SharedPreferences 中的 NVP。get<Type>()函数的第 1 个参数是 NVP 的名称，第二个参数是默认值，在无法获取数值时使用。

```
SharedPreferences   sharedPreferences= getSharedPreferences ( PREFERENCE
NAME, MODE) ,
String    name= sharedPreferences .getString ( "Name ", "Default Name " );
int    age= sharedPreferences .getInt ( "Age ", 20) ;
float    height= sharedPreferences . getFloat ( "Height", 1. 8lf);
```

8.1.2　SharedPreferences 应用（【例一】：存取注册信息）

【例一】本实例由两个 Android 工程组成，两个工程分别单独创建。工程 SharedReg 实现注册功能，在界面上输入用户信息后提交，将用户填写的信息用 SharedPreferences 存储。工程 SharedLog 实现登录功能，初启动时就用 SharedPreferences 读入注册的用户信息，用户名称显示在"用户名"栏，用户输入密码后，程序将其与 SharedPreferences 存储的密码进行核对验证，若正确显示欢迎页。错误则给出消息提示。

1. 创建注册应用框架

创建 Android 工程 SharedReg，采用第 5 章 RegisterPage 工程注册页面输入注册信息，单击"提交"，将注册的用户信息以 SharedPreferences 方式保存起来。如图 8.1 和图 8.2 所示。

图 8.1　注册页面　　　　　　图 8.2　注册成功页面

（1）界面复用第 5 章 RegisterPage 工程的 register.xml、success.xml 文件，将它们复制到本工程 res\layout 下。

（2）activity_main.xml 文件内容修改简化（即去除工具栏）如下：

```xml
<?xml version="1.0" encoding="utf-8"?>
<android.support.design.widget.CoordinatorLayout xmlns:android="http://schemas.android.com/apk/res/android"
    xmlns:app="http://schemas.android.com/apk/res-auto"
    xmlns:tools="http://schemas.android.com/tools"
    android:layout_width="match_parent"
    android:layout_height="match_parent"
    android:fitsSystemWindows="true"
    tools:context="com.easybooks.android.sharedreg.MainActivity">

    <include layout="@layout/register" />
```

```xml
<android.support.design.widget.FloatingActionButton
    android:id="@+id/fab"
    android:layout_width="wrap_content"
    android:layout_height="wrap_content"
    android:layout_gravity="bottom|end"
    android:layout_margin="@dimen/fab_margin"
    android:src="@android:drawable/ic_dialog_email" />

</android.support.design.widget.CoordinatorLayout>
```

(3) MainActivity.java 代码修改简化，去除工具栏，onCreate()仅仅保留两行，具体代码如下：

```java
package com.easybooks.android.sharedreg;
…
public class MainActivity extends AppCompatActivity {
    @Override
    protected void onCreate(Bundle savedInstanceState) {
        super.onCreate(savedInstanceState);
        setContentView(R.layout.activity_main);
    }
    …
}
```

(4) 在工程中创建一个 User 类，采用 SharedPreference，实现用户信息的持久化存取功能。主要包括以下几方面。

① 声明一个 SharedPreferences 对象用于保存数据。

② 设计 User 类方法 saveUser()，以名/值方式将用户注册 User 属性数据保存起来。

③ 设计 User 类方法 getUserData()，以名/值方式读取注册用户一组属性数据。

User.java 代码如下：

```java
package com.easybooks.android.sharedreg;

import android.content.Context;
import android.content.SharedPreferences;

import java.io.Serializable;

public class User implements Serializable {
    private String name;
    private String pwd;
    private int sex;
    private String birth;
    private String degree;
    //① 声明一个 SharedPreferences 对象用于保存数据
    private SharedPreferences sharedPreferences = null;
    private static final String SHARED_NAME = "SharedUserInfo";    //记录用户信息的文件名
    private static int MODE = Context.MODE_WORLD_READABLE + Context.MODE_WORLD_WRITEABLE;        //权限一定要设为允许读写

    public User() {
    }
```

```java
public User(String name, String pwd, int sex, String birth, String degree) {
    this.name = name;
    this.pwd = pwd;
    this.sex = sex;
    this.birth = birth;
    this.degree = degree;
}

@Override
public String toString() {
    String userInfo = name + " " + ((sex == 1) ? "先生" : "女士") + ", 恭喜您注册成功！\n 您的注册信息为：\n 出生日期         " + birth + "\n 学        历        " + degree;
    return userInfo;
}
//② 以名/值方式将用户注册一组属性保存起来
public void saveUser(Context context) {
    //创建 SharedPreferences 对象
    sharedPreferences = context.getSharedPreferences(SHARED_NAME, MODE);
    //得到 Editor 对象
    SharedPreferences.Editor editor = sharedPreferences.edit();
    //保存数据
    editor.putString("name", name);
    editor.putString("pwd", pwd);
    editor.putInt("sex", sex);
    editor.putString("birth", birth);
    editor.putString("degree", degree);
    editor.commit();                        //提交
}
//③ 以名/值方式读取注册用户一组属性信息
public void getUserData(Context context) {
    //得到 SharedPreferences 对象
    sharedPreferences = context.getSharedPreferences(SHARED_NAME, MODE);
    //获取注册的用户数据
    name = sharedPreferences.getString("name", "");
    pwd = sharedPreferences.getString("pwd", "");
    sex = sharedPreferences.getInt("sex", 1);
    birth = sharedPreferences.getString("birth", "");
    degree = sharedPreferences.getString("degree", "");
}
}
```

（5）修改第 5 章 RegisterPage 工程中 RegisterActivity.java，把用户界面输入的注册信息通过 SharedPreferences 形式保存起来。

RegisterActivity.java 代码如下：

```java
package com.easybooks.android.sharedreg;
…
public class RegisterActivity extends AppCompatActivity {
    …
    public void onSubmitClick(View view) {
        String name = myName.getText().toString();
        String pwd = myPwd.getText().toString();
```

```
        int sex = Integer.parseInt(mySex.getText().toString().equals("男") ? "1" : "0");
        String birth = myBirth.getText().toString();
        String degree = mydegreeTemp;
        User user = new User(name, pwd, sex, birth, degree);    //①
        user.saveUser(this);                                    //②
        //跳转页面
        Intent intent = new Intent(this, SuccessActivity.class);
        startActivity(intent);
    }
}
```

其中:

① 创建自定义的 User 对象存储用户信息。

② 调用其 saveUser 方法来保存用户注册的个人信息。

(6) 编写 SuccessActivity.java，代码如下。

```
package com.easybooks.android.sharedreg;
    ...
public class SuccessActivity extends AppCompatActivity {
    private TextView mySuccess;
    private Button myEnter;
    @Override
    protected void onCreate(Bundle savedInstanceState) {
        super.onCreate(savedInstanceState);
        setContentView(R.layout.success);
        findViews();
        showSuccess();
    }
    private void findViews() {
        mySuccess = (TextView) findViewById(R.id.myLabelSuccess);
        myEnter = (Button) findViewById(R.id.myButtonEnter);
    }

    private void showSuccess() {
        User user = new User();
        user.getUserData(this);                         //①
        mySuccess.setText(user.toString());             //②
    }
}
```

其中:

① 把调用 getUserData 方法获取 SharedPreferences 形式保存的用户注册信息保存到 User 类属性中。

② 将 User 类属性信息显示出来。

由此可见，这里将数据存取操作的实现细节完全封装于自定义的 User 类中，本章所有的实例都将复用同一个注册应用框架来演示 Android 中各种不同的数据存储方式，只需修改 User 类的代码即可，外部 RegisterActivity 和 SuccessActivity 的代码完全不要动，直接将文件复制过去，修改命名空间即可。

（7）配置 AndroidManifest.xml 如下。

```xml
<?xml version="1.0" encoding="utf-8"?>
<manifest xmlns:android="http://schemas.android.com/apk/res/android"
    package="com.easybooks.android.sharedreg">

    <application
    …
        <activity
            android:name=".RegisterActivity"
            android:label="@string/app_name"
            android:theme="@style/AppTheme.NoActionBar">
            <intent-filter>
                <action android:name="android.intent.action.MAIN" />

                <category android:name="android.intent.category.LAUNCHER" />
            </intent-filter>
        </activity>
        <activity android:name=".SuccessActivity"/>
    </application>

</manifest>
```

2. 创建登录应用框架

创建 Android 工程 SharedLog。采用第 5 章例一（LoginPage 工程）登录页面输入用户名和密码，单击"登录"，将用户信息与用 SharedPreferences 方式保存的注册信息的名称和密码进行对比。不一致，显示"用户名和密码错误"提示，否则显示登录成功的欢迎页。如图 8.3 和图 8.4 所示。

图 8.3　输入密码错误　　　　　　　　图 8.4　登录成功的欢迎页

（1）界面设计复用第 5 章例一（LoginPage 工程）的 login.xml、welcome.xml 文件，将它们复制到本项目 res\layout 下，将图片资源 androidwelcomer.gif 复制到 res\drawable 目录下。

（2）activity_main.xml、MainActivity.java 文件的内容均加以简化，去除工具栏元素，方法同前。

在 activity_main.xml 中设 login.xml 为启动页：

```
tools:context="com.easybooks.android.sharedlog.MainActivity">
<include layout="@layout/login" />
```

（3）编写 LoginActivity.java，包括以下几方面。

① 先定义变量 SharedPreferences 方式：名称、读和写工作方式。

② 获取 SharedPreferences 存储数据用户名和密码并保存到全局变量中。

③ 单击"登录"按钮，如果当前界面输入登录名和密码与注册信息相同，显示欢迎页面，否则提示错误信息。

LoginActivity.java 代码如下：

```java
package com.easybooks.android.sharedlog;

import android.content.Context;
…

public class LoginActivity extends AppCompatActivity {
    private EditText myName;
    private EditText myPwd;

    //① 声明相关的全局变量
    private final static String SHARED_PACKAGE = "com.easybooks.android.sharedreg";
                            //外部应用所在的包
    private static final String SHARED_NAME = "SharedUserInfo";
                            //记录用户信息的文件名
    private Context context = null;    //声明 Context 对象用于保存外部应用的上下文
    private SharedPreferences sharedPreferences = null;
    private static int MODE = Context.MODE_WORLD_READABLE + Context.MODE_WORLD_WRITEABLE;
                            //权限一定要设为允许读写
    private String myusername;         //用户名
    private String mypassword;         //密码

    @Override
    protected void onCreate(Bundle savedInstanceState) {
        super.onCreate(savedInstanceState);
        setContentView(R.layout.login);
        findViews();
        loadUser();              //②
    }

    private void findViews() {
        myName = (EditText) findViewById(R.id.myTextName);
        myPwd = (EditText) findViewById(R.id.myTextPwd);
    }
    //② 获得 SharedPreferences 数据并把用户名和密码存储到全局变量中
    private void loadUser() {
        try {
            context = createPackageContext(SHARED_PACKAGE, CONTEXT_IGNORE_SECURITY);
            //得到注册应用的 SharedPreferences 对象
            sharedPreferences = context.getSharedPreferences(SHARED_NAME, MODE);
            //获取已注册的用户名和密码并加载显示在界面上
            myusername = sharedPreferences.getString("name", "");
            myName.setText(myusername);
            mypassword = sharedPreferences.getString("pwd", "");
```

```java
        } catch (PackageManager.NameNotFoundException e) {
            e.printStackTrace();
        }
    }
    //③ 单击"登录"按钮
    public void onLoginClick(View view) {
        String name = myName.getText().toString();
        String pwd = myPwd.getText().toString();
        //判断当前界面输入登录名和密码与注册信息是否相同
        if (name.equals(myusername) && pwd.equals(mypassword)) {
            Intent intent = new Intent(this, WelcomeActivity.class);
            Bundle bundle = new Bundle();
            bundle.putString("name", name);
            intent.putExtras(bundle);
            startActivity(intent);
        } else {
            Toast.makeText(this, "用户名或密码错！请重输。", Toast.LENGTH_SHORT).show();
            myName.setText(myusername);
            myPwd.setText("");
        }
    }
}
```

（4）编写 WelcomeActivity.java，代码如下。

```java
package com.easybooks.android.sharedlog;
…
public class WelcomeActivity extends AppCompatActivity {
    private TextView myWelcome;

    @Override
    protected void onCreate(Bundle savedInstanceState) {
        super.onCreate(savedInstanceState);
        setContentView(R.layout.welcome);
        findViews();
        showWelcome();
    }

    private void findViews() {
        myWelcome = (TextView) findViewById(R.id.myLabelWelcome);
    }

    private void showWelcome() {
        Bundle bundle = getIntent().getExtras();
        String name = bundle.getString("name");
        myWelcome.setText("\n" + name + " 您好！\n    欢迎光临");
    }

    public void onBackClick(View view) {
        finish();
    }
}
```

说明：

这里对数据的存取操作都集中于 LoginActivity 的 loadUser 方法中，本章所有的实例也都将复用这同一个登录应用框架来演示 Android 中不同存储方式下数据的跨应用共享，只须重写 loadUser 的代码并修改对应全局变量的声明语句即可，而 WelcomeActivity 及其他源文件的代码完全不要动，直接将文件复制过去用。

（5）配置 AndroidManifest.xml 如下。

```xml
<?xml version="1.0" encoding="utf-8"?>
<manifest xmlns:android="http://schemas.android.com/apk/res/android"
    package="com.easybooks.android.sharedlog">

    <application
     …
        <activity
            android:name=".LoginActivity"
            android:label="@string/app_name"
            android:theme="@style/AppTheme.NoActionBar">
            <intent-filter>
                <action android:name="android.intent.action.MAIN" />

                <category android:name="android.intent.category.LAUNCHER" />
            </intent-filter>
        </activity>
        <activity android:name=".WelcomeActivity"/>
    </application>

</manifest>
```

3. 运行效果

先打开工程 SharedReg 运行，输入注册信息后提交。结束运行，关闭本工程。

打开本例的另一个工程 SharedLog 运行。可以看到，程序在初始启动的界面上就已经自动加载显示出了刚刚注册的用户名，这个用户名是由前一个应用 SharedReg 通过 SharedPreferences 存储到共享目录的，故应用 SharedLog 也能访问到。

当用户输入密码登录时，程序会与之前注册时填写的密码核对，若不一致，自动清空密码栏并提示用户重新输入。若正确，则跳转到欢迎页。

4. 观察存储的数据文件

用共享优先方式存储的数据文件保存在模拟器/data/data/<项目包名>/shared_prefs 目录下，打开工程 SharedReg，启动模拟器。

在 Android Studio 内，选择主菜单 Tools|Android|Android Device Monitor，打开模拟器的文件管理器，在相应的路径下可找到这个文件，如图 8.5 所示。

点击图 8.5 窗口右上角的"Pull a file from the device"（图标为）按钮，可将该文件导出并保存至本地计算机硬盘，以记事本打开可查看到其中的内容（正是用户刚刚注册时输入的信息），如图 8.6 所示。

第 8 章　Android 数据存储与共享

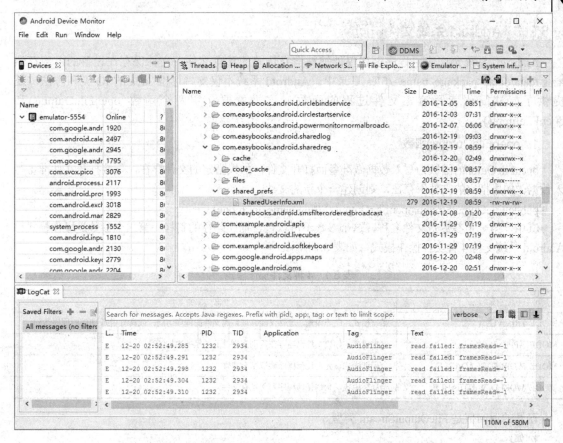

图 8.5　用 SharedPreferences 存储到共享目录的数据文件

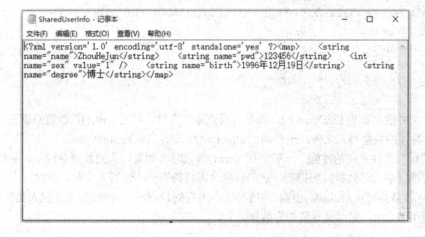

图 8.6　SharedPreferences 存储文件的内容

8.2　内部文件存储

　　虽然 SharedPreferences 能够为开发人员简化数据存储和访问过程，但直接使用文件系统保存数据仍然是 Android 数据存储中不可或缺的组成部分。Android 使用 Linux 的文件系统，开发人员可以建立和访问程序自身建立的私有文件。

8.2.1 Android 系统文件访问

Android 系统允许应用程序创建仅能够被自身访问的私有文件,文件保存在设备的内部存储器上/data/data/<package name>/files 目录中。Android 系统不仅支持标准 Java 的 IO 类和方法,还提供了能够简化读写流式文件过程的函数。这里主要介绍两个函数 openFileOutput()和 openFileInput()。

1. openFileOutput()函数

openFileOutput()函数为写入数据做准备而打开文件。如果指定的文件存在,直接打开文件准备写入数据。如果指定的文件不存在,则创建一个新的文件。

public FileOutputStream openFileOutput (String name, int mode)

其中,第一个参数是文件名称,这个参数不可以包含描述路径的斜杠。第二个参数是操作模式,Android 系统支持 4 种文件操作模式,如表 8.1 所示。

表 8.1 4 种文件操作模式

模 式	说 明
MODE_PRIVATE	私有模式,缺陷模式,文件仅能够被创建文件的程序访问,或具有相同 UID 的程序访问
MODE_APPEND	追加模式,如果文件已经存在,则在文件的结尾处添加新数据
MODE_WORLD_READABLE	全局读模式,允许任何程序读取私有文件
MODE_WORLD_WRITEABLE	全局写模式,允许任何程序写入私有文件

函数的返回值是 FileOutputStream 类型。

例如:

```
String FILE_NAME = "fileDemo.txt";
FileOutputStream fos = openFileOutput(FILE_NAME,Context.MODE_PRIVATE)
String text = "Some data";
fos.write(text.getBytes());
fos.flush();
fos.close();
```

其中:

(1)定义文件的名称为 fileDemo.txt,以私有模式建立文件,调用 write()函数将数据写入文件,调用 flush()函数将缓冲数据写入文件,最后调用 close()函数关闭 FileOutputStream。

(2)为了提高文件系统的性能,一般调用 write()函数时,如果写入的数据量较小,系统会把数据保存在数据缓冲区中,等数据量积攒到一定的程度时再将数据一次性写入文件。因此,在调用 close()函数关闭文件前,务必要调用 flush()函数,将缓冲区内所有的数据写入文件。如果开发人员在调用 close()函数前没有调用 flush(),则可能导致部分数据丢失。

2. openFileInput()函数

openFileInput()函数为读取数据做准备而打开文件。

public FileInputStream openFileInput (String name)

其中,第一个参数也是文件名称,同样不允许包含描述路径的斜杠。

例如,以二进制方式读取数据。

```
String FILE_NAME = "fileDemo.txt";
FileInputStream fis = openFileInput(FILE_NAME);
```

```
byte[] readBytes = new byte[fis.available()];
while(fis.read(readBytes) != -1){
}
```

上面的两部分代码在实际使用过程中会遇到错误提示，这是因为文件操作可能会遇到各种问题而最终导致操作失败，因此在代码中应使用 try/catch 捕获可能产生的异常。

8.2.2 文件存储应用（【例二】：存取注册信息）

【例二】用内部文件来保存用户注册的信息，采用例一的程序框架，由于文件方式存储的数据只能由创建它的程序自身访问，不支持跨应用访问，故本例只包含一个 Android 工程，实现注册及信息回显的功能。

1. 创建注册应用

创建 Android 工程 InterReg，直接套用例一开发好的注册应用框架，将相关的源文件复制到本工程对应的目录即可。

2. 实现文件存取

需要修改重新实现的只有 User 类，本例改用内部文件的方式来实现存取。

User.java 代码如下：

```java
package com.easybooks.android.interreg;

import android.content.Context;

import java.io.FileInputStream;
import java.io.FileNotFoundException;
import java.io.FileOutputStream;
import java.io.IOException;
import java.io.Serializable;

public class User implements Serializable {
    private String name;
    private String pwd;
    private int sex;
    private String birth;
    private String degree;
    //声明 FileOutputStream/FileInputStream 流类用于访问手机内存
    private FileOutputStream fos = null;
    private FileInputStream fis = null;
    private static final String INTER_NAME = "InterUserInfo";    //记录用户信息的文件名
    private static int MODE = Context.MODE_WORLD_READABLE + Context.MODE_WORLD_WRITEABLE;                                     //权限一定要设为允许读写

    public User() {
    }

    public User(String name, String pwd, int sex, String birth, String degree) {
        this.name = name;
        this.pwd = pwd;
        this.sex = sex;
```

```java
            this.birth = birth;
            this.degree = degree;
        }

        @Override
        public String toString() {
            String userInfo = name + " " + ((sex == 1) ? "先生" : "女士") + ", 恭喜您注册成功! \n 您的注册信息为:
\n 出生日期        " + birth + "\n 学        历        " + degree;
            return userInfo;
        }

        public void saveUser(Context context) {
            try {
                //获得文件输出流
                fos = context.openFileOutput(INTER_NAME, MODE);
                //保存数据
                String userdata = name;
                userdata += ";" + pwd;
                userdata += ";" + String.valueOf(sex);
                userdata += ";" + birth;
                userdata += ";" + degree;
                fos.write(userdata.getBytes());
                fos.flush();                           //清除缓存
            } catch (FileNotFoundException e) {
                e.printStackTrace();
            } catch (IOException e) {
                e.printStackTrace();
            } finally {
                if (fos != null) {
                    try {
                        fos.close();                   //及时关闭文件输出流
                    } catch (IOException e) {
                        e.printStackTrace();
                    }
                }
            }
        }

        public void getUserData(Context context) {
            try {
                //获得文件输入流
                fis = context.openFileInput(INTER_NAME);
                //定义暂存数据的数组
                byte[] buffer = new byte[fis.available()];
                //从输入流中读取数据
                fis.read(buffer);
                //获取注册的用户数据
                String data = new String(buffer);
                name = data.split(";")[0];
                pwd = data.split(";")[1];
                sex = Integer.parseInt(data.split(";")[2]);
                birth = data.split(";")[3];
                degree = data.split(";")[4];
```

```
            } catch (FileNotFoundException e) {
                e.printStackTrace();
            } catch (IOException e) {
                e.printStackTrace();
            } finally {
                if (fis != null) {
                    try {
                        fis.close();          //及时关闭文件输入流
                    } catch (IOException e) {
                        e.printStackTrace();
                    }
                }
            }
        }
    }
}
```

3. 运行效果

运行效果同例一的注册应用，见图 8.1 和图 8.2。

4. 观察存储的数据文件

用内部文件方式存储的数据文件保存在模拟器/data/data/<项目包名>/files 目录下，打开工程 InterReg，启动模拟器。

在 Android Studio 内，选择主菜单 Tools|Android|Android Device Monitor，打开模拟器的文件管理器，在相应的路径下可找到这个文件，如图 8.7 所示。

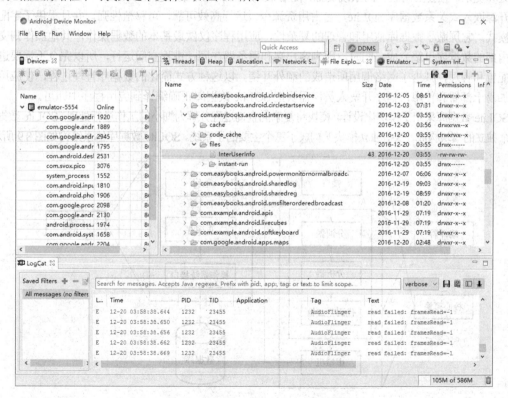

图 8.7　用内部文件方式存储的数据文件

点击窗口右上角图标按钮，可将该文件导出并保存至本地计算机硬盘，以记事本打开可查看到其中的内容（也正是用户刚刚注册时输入的信息），如图 8.8 所示。

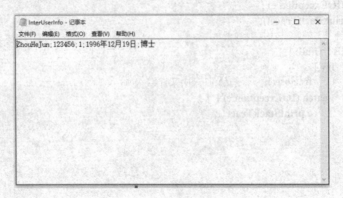

图 8.8　内部文件的内容

8.3　SQLite 数据库存储与共享

8.3.1　SQLite 概述

自从出现商业应用程序以来，数据库就一直是应用程序的主要组成部分，数据库的管理系统也比较庞大和复杂，且会占用较多的系统资源。随着嵌入式应用程序的大量出现，2000 年，D.Richard Hipp 开发开源嵌入式关系数据库 SQLite，它占用资源少，运行高效可靠，可移植性强，并且提供了零配置运行模式。它屏蔽了数据库使用和管理的复杂性，应用程序仅做最基本的数据操作，其他操作则交给进程内部的数据库引擎完成。同时，因为客户端和服务器在同一进程空间运行，所以完全不需要进行网络配置和管理，减少了网络调用所造成的额外开销。以这种方式简化的数据库管理过程，使应用程序更加易于部署和使用，程序开发人员仅需要把 SQLite 数据库正确编译到应用程序中即可。

SQLite 数据库采用了模块化设计，模块将复杂的查询过程分解为细小的工作进行处理。SQLite 数据库由 8 个独立的模块构成，这些独立模块又构成了三个主要的子系统。SQLite 数据库体系结构如图 8.9 所示。

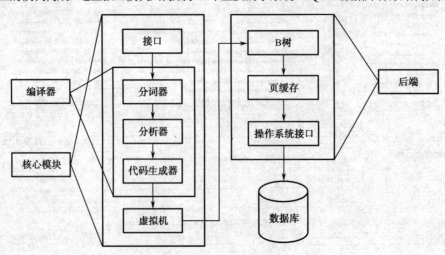

图 8.9　SQLite 数据库体系结构

接口由 SQLite C API 组成，因此无论是应用程序、脚本，还是库文件，最终都是通过接口与 SQLite 交互。

在编译器中，分词器和分析器对 SQL 语句进行语法检查，然后把 SQL 语句转化为便于底层处理的分层数据结构，这种分层的数据结构称为"语法树"。然后把语法树传给代码生成器进行处理，生成一种用于 SQLite 的汇编代码，最后由虚拟机执行。

SQLite 数据库体系结构中最核心的部分是虚拟机，也称为虚拟数据库引擎(Virtual Database Engine，VDBE)。与 Java 虚拟机相似，虚拟数据库引擎用来解释并执行字节代码。虚拟数据库引擎的字节代码由 128 个操作码构成，这些操作码主要用以对数据库进行操作，每一条指令都可以完成特定的数据库操作，或以特定的方式处理栈的内容。

后端由 B-树、页缓存和操作系统接口构成，B-树和页缓存共同对数据进行管理。B-树的主要功能就是索引，它维护着各个页面之间复杂的关系，便于快速找到所需数据。页缓存的主要作用是通过操作系统接口在 B-树和磁盘之间传递页面。

SQLite 数据库具有很强的移植性，可以运行在 Windows、Linux、BSD、Mac OS 和一些商用 Unix 系统中，比如 Sun 的 Solaris 或 IBM 的 AIX。同样，也可以工作在许多嵌入式操作系统下，比如 QNX、VxWorks、Palm OS、Symbian 和 Windows CE。SQLite 的核心大约有 3 万行标准 C 代码，因为模块化的设计使这些代码非常易于理解。

8.3.2　SQLite 应用（【例三】：存取注册信息）

用 SQLite 数据库来保存用户注册的信息，依然采用例一的程序框架。由于 SQLite 方式存储的数据 Android 官方建议使用 ContentProvider 实现与其他应用的共享，本例先开发一个 Android 项目来实现注册回显功能。稍后在介绍了 ContentProvider 后，运用 ContentProvider 将数据库中的表数据对外公开（通过 URI 地址），再开发一个登录应用去访问数据库的数据。

【例三】SQLite 数据库存取注册信息。

1. 创建注册应用

创建 Android 工程 SqliteReg，直接套用例一开发好的注册应用框架，将相关的源文件复制到本项目对应的目录即可。

2. 创建 SQLite 适配器类

在工程中创建 SQLiteAdapter 类用于封装 SQLite 的创建、打开、关闭及获取数据库实例的操作。SQLiteAdapter.java 代码如下：

```
package com.easybooks.android.sqlitereg;

import android.content.Context;
import android.database.sqlite.SQLiteDatabase;
import android.database.sqlite.SQLiteOpenHelper;

public class SQLiteAdapter {
    //声明数据库的基本信息
    private static final int DB_VERSION = 1;                    //数据库版本
    private static final String TABLE_NAME = "users";           //记录用户信息的表名
    //表中各个字段定义
    private static final String _ID = "_id";                    //保存 ID 值
    private static final String NAME = "name";                  //用户名
    private static final String PWD = "pwd";                    //密码
```

```java
        private static final String SEX = "sex";              //性别
        private static final String BIRTH = "birth";          //出生日期
        private static final String DEGREE = "degree";        //学历
    //声明操作Sqlite数据库的实例
    private SQLiteDatabase sqliteDb;
    private DBOpenHelper sqliteHelper;

    //构造方法
    public SQLiteAdapter(Context context, String dbname) {
        sqliteHelper = new DBOpenHelper(context, dbname, null, DB_VERSION);
        sqliteDb = sqliteHelper.getWritableDatabase();       //获得可写的数据库
    }

    //自定义的帮助类
    private static class DBOpenHelper extends SQLiteOpenHelper {
        public DBOpenHelper(Context context, String dbname, SQLiteDatabase.CursorFactory factory, int version) {
            super(context, dbname, factory, version);
        }

        private static final String CREATE_TABLE = "create table " + TABLE_NAME
                + "(" + _ID + " integer primary key autoincrement,"
                + NAME + " text not null,"
                + PWD + " text not null,"
                + SEX + " integer not null,"
                + BIRTH + " text,"
                + DEGREE + " text);";                        //预定义创建表的SQL语句

        @Override
        public void onCreate(SQLiteDatabase db) {
            db.execSQL(CREATE_TABLE);                        //创建表
        }

        @Override
        public void onUpgrade(SQLiteDatabase db, int oldVersion, int newVersion) {
            db.execSQL("drop table if exists " + TABLE_NAME);
                                                             //删除老表
            onCreate(db);                                    //创建新表
        }
    }

    //获取SQLite数据库实例
    public SQLiteDatabase getSqliteDb() {
        return sqliteDb;
    }
}
```

3. 用 SQLite 保存用户信息

修改重新实现 User 类，本例改用 SQLite 来实现存取。
User.java 代码如下：

```java
package com.easybooks.android.sqlitereg;

import android.content.ContentValues;
```

```java
import android.content.Context;
import android.database.Cursor;
import android.database.sqlite.SQLiteDatabase;

import java.io.Serializable;


public class User implements Serializable {
    private int _id;//保存用户的 ID(如果计划使用 ContentProvider 来共享表,就必须具有唯一的 ID 字段)
    private String name;
    private String pwd;
    private int sex;
    private String birth;
    private String degree;
    //声明一个 SQLiteAdapter 对象作为访问 SQLite 数据库的中介
    private SQLiteAdapter sqLiteAdapter = null;
    private SQLiteDatabase sqLiteDb = null;                      //数据库实例
    private static final String SQLITE_NAME = "SqliteUserInfo";  //数据库名称
    private static final String TABLE_NAME = "users";            //记录用户信息的表名

    public User() {
    }

    public User(String name, String pwd, int sex, String birth, String degree) {
        this.name = name;
        this.pwd = pwd;
        this.sex = sex;
        this.birth = birth;
        this.degree = degree;
    }

    @Override
    public String toString() {
        String userInfo = name + " " + ((sex == 1) ? "先生" : "女士") + ", 恭喜您注册成功!\n 您的注册信息为:
\n 出生日期           " + birth + "\n 学       历         " + degree;
        return userInfo;
    }

    public void saveUser(Context context) {
        sqLiteAdapter = new SQLiteAdapter(context, SQLITE_NAME);
                                                        //创建 SQLiteAdapter 对象
        sqLiteDb = sqLiteAdapter.getSqliteDb();         //得到 SQLite 实例
           ContentValues values = new ContentValues();  //构造 ContentValues 实例
        //保存数据
        values.put("name", name);
        values.put("pwd", pwd);
        values.put("sex", sex);
        values.put("birth", birth);
        values.put("degree", degree);
        sqLiteDb.insert(TABLE_NAME, null, values);      //添加数据
    }

    public void getUserData(Context context) {
```

```
            sqLiteAdapter = new SQLiteAdapter(context, SQLITE_NAME);
                                                        //创建 SQLiteAdapter 对象
            sqLiteDb = sqLiteAdapter.getSqliteDb();     //得到 SQLite 实例
            Cursor cursor = sqLiteDb.query(TABLE_NAME, new String[]{"name", "pwd", "sex", "birth", "degree"}, null, null, null, null, null);
                                                        //获取注册的用户数据
            if (cursor.getCount() > 0) {
                cursor.moveToFirst();
                name = cursor.getString(0);
                pwd = cursor.getString(1);
                sex = cursor.getInt(2);
                birth = cursor.getString(3);
                degree = cursor.getString(4);
            }
            cursor.close();
        }
    }
```

4. 运行效果

运行效果同例一的注册应用，见图 8.1 和图 8.2。

5. 观察存储的数据

用 SQLite 数据库方式存储的数据文件保存在模拟器/data/data/<项目包名>/databases 目录下，打开工程 SqliteReg，启动模拟器。

在 Android Studio 内，选择主菜单 Tools|Android|Android Device Monitor，打开模拟器的文件管理器，在相应的路径下可找到这个文件，如图 8.10 所示。

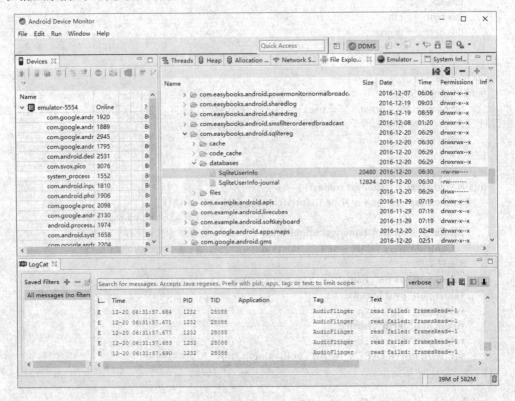

图 8.10 用 SQLite 方式存储的数据文件

查看 SQLite 数据库表的内容，一般通过 Android 系统自带的 sqlite3 进行，在 Android SDK 的 tools 目录中有 sqlite3 工具，可通过 Windows 命令行启动。操作前要确保已经配置了正确的环境变量，以便 Windows 能够找到 sqlite3，如图 8.11 所示。

图 8.11 配置环境变量

打开 Windows 命令行，输入 adb shell 命令，进入本项目目录/data/data/com.easybooks.android.sqlitereg/databases。

输入 ls 回车，可看到项目运行所生成的 SQLite 数据库文件 SqliteUserInfo。

输入 sqlite3 SqliteUserInfo 回车，启动 sqlite3 工具。

输入 .tables 回车，查看数据库中的表，这时可看到我们程序所创建的 users 表，该表中存有注册用户的信息，用 select 语句直接查询即可看到内容，整个过程如图 8.12 所示。

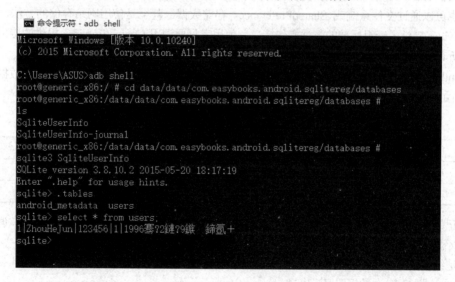

图 8.12 查看 SQLite 表数据

8.4 ContentProvider 数据共享组件

8.4.1 ContentProvider 组件

ContentProvider 是应用程序之间共享数据的一种接口机制。应用程序运行在不同的进程中，因此数据和文件在不同的应用程序之间是不能够直接访问的。通过 ContentProvider，应用程序可以指定需要共享的数据，而其他应用程序则可以在不知道数据来源、路径的情况下，对共享数据进行查询、添加、删除和更新等操作。

在创建 ContentProvider 前，首先要实现底层的数据源。数据源包括数据库、文件系统或网络等，然后继承 ContentProvider 类中实现基本数据操作的接口函数，包括添加、删除、查找和更新等功能。调用者不能直接调用 ContentProvider 的接口函数，而需要使用 ContentResolver 对象，通过 URI 间接调用 ContentProvider，调用关系如图 8.13 所示。

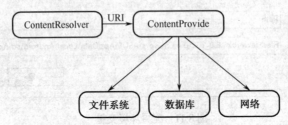

图 8.13 ContentProvider 调用关系

在 ContentResolver 对象与 ContentProvider 进行交互时，通过 URI 确定要访问的 ContentProvider 数据集。在发起一个请求的过程中，系统根据 URI 确定处理这个查询的 ContentProvider，然后系统初始化 ContentProvider 所有需要的资源。一般情况下只有一个 ContentProvider 对象，但却可以同时与多个 ContentResolver 进行交互。

ContentProvider 完全屏蔽了数据提供组件的数据存储方法。数据提供者可以使用 SQLite 数据库存储数据，也可以通过文件系统或 SharedPreferences 存储数据，甚至是使用网络存储的方法，它通过 ContentProvider 提供了一组标准的数据操作接口，使用者只要调用 ContentProvider 提供的接口函数，即可完成所有的数据操作。

ContentProvider 的数据集类似于数据库的数据表，每行是一条记录，每列具有相同的数据类型，如表 8.2 所示。

表 8.2 ContentProvider 数据集

_ID	NAME	AGE	HEIGHT
1	Tom	21	1.81
2	Jim	22	1.78

每条记录都包含一个长型的字段_ID，用来唯一标识每条记录。可以提供多个数据集，调用者使用 URI 对不同的数据集的数据进行操作。

URI 是统一资源标识符，用来定位远程或本地的可用资源。

content ://<授权者>/<数据路径> /< id>

其中：

content://：表示该 URI 用于 ContentProvider 定位资源。

<授权者>：确定 ContentProvider 资源。一般由类的小写全称组成，以保证唯一性。

<数据路径>：确定数据集。如果 ContentProvider 仅提供一个数据集，数据路径则是可以省略的。如果提供多个数据集，数据路径则必须指明具体是哪一个数据集。数据集的数据路径可以写成多段格式。

<id>是数据编号，用来唯一确定数据集中的一条记录，用来匹配数据集中_ID 字段的值。如果请求的数据并不只限于一条数据，则<id>是可以省略的。

例如：

请求整个 computer 数据集的 URI 应写为

content : //edu.njone.computerprovider/computer

请求 computer 数据集中第三条数据的 URI 则应写为

content : //edu.njone.computerprovider/computer/3

8.4.2 ContentProvider 创建

通过继承 ContentProvider 类可以创建一个新的数据提供者，过程可以分为三步。

1. 继承 ContentProvider 重载函数

新建立的类继承 ContentProvider 后，delete()、insert()、query()、update()和 onCreate()、getType()函数需要重载。其中，前面 4 个分别用于对数据集的删除、添加、查询和更新操作。onCreate()一般用来初始化底层数据集和建立数据连接等工作。getType()函数用来返回指定 URI 的 MIME 数据类型。如果 URI 是单条数据，则返回的 MIME 数据类型应以 vnd.android.cursor.item/开头。如果 URI 是多条数据，则返回的 MIME 数据类型应以 vnd.android.cursor.dir/开头。

新建立的类继承 ContentProvider 后，Eclipse 会提示程序开发人员需要重载部分的代码，并自动生成需要重载的代码框架。

下面的代码是 Eclipse 自动生成的代码框架。

```
import android.content.*;
import android.database.Cursor;
import android.net.Uri;

public class PeopleProvider extends ContentProvider{
    @Override
    public int delete(Uri uri, String selection, String[] selectionArgs) {
        // TODO Auto-generated method stub
        return 0;
    }
    @Override
    public String getType(Uri uri) {
        // TODO Auto-generated method stub
        return null;
    }
    @Override
    public Uri insert(Uri uri, ContentValues values) {
        // TODO Auto-generated method stub
        return null;
    }

    @Override
    public boolean onCreate() {
        // TODO Auto-generated method stub
        return false;
    }
    @Override
    public Cursor query(Uri uri, String[] projection, String selection,
            String[] selectionArgs, String sortOrder) {
        // TODO Auto-generated method stub
```

```
                return null;
            }
            @Override
            public int update(Uri uri, ContentValues values, String selection,
                    String[] selectionArgs) {
                // TODO Auto-generated method stub
                return 0;
            }
        }
```

2. 声明 CONTENT_URI 实现 UriMatcher

在新构造的 ContentProvider 类中，需构造一个 UriMatcher 判断 URI 是单条数据还是多条数据。同时，为了便于判断和使用 URI，一般将 URI 的授权者名称和数据路径等内容声明为静态常量，并声明 CONTENT_URI。

声明 CONTENT_URI 和构造 UriMatcher 的代码如下：

```
public static final String AUTHORITY = "edu.hrbeu.peopleprovider";                //(1)
public static final String PATH_SINGLE = "people/#";                              //(2)
public static final String PATH_MULTIPLE = "people";                              //(2)
public static final String CONTENT_URI_STRING = "content://" + AUTHORITY + "/" + PATH_MULTIPLE;
                                                                                  //(3)
public static final Uri CONTENT_URI = Uri.parse(CONTENT_URI_STRING);              //(4)
private static final int MULTIPLE_PEOPLE = 1;                                     //(5)
private static final int SINGLE_PEOPLE = 2;                                       //(5)

private static final UriMatcher uriMatcher;                                       //(6)
static {
    uriMatcher = new UriMatcher(UriMatcher.NO_MATCH);
    uriMatcher.addURI(AUTHORITY, PATH_SINGLE, MULTIPLE_PEOPLE);
    uriMatcher.addURI(AUTHORITY, PATH_MULTIPLE, SINGLE_PEOPLE);
}
```

其中：

（1）声明 URI 的授权者名称。
（2）声明单条和多条数据的数据路径。
（3）声明 CONTENT_URI 的字符串形式。
（4）正式声明 CONTENT_URI。
（5）声明多条数据和单条数据的返回代码。
（6）声明 UriMatcher，并在静态构造函数中声明匹配方式和返回代码。

其中，UriMatcher.NO_MATCH 是 URI 无匹配时的返回代码。public void addURI (String authority, String path, int code)函数用来添加新的匹配项。authority 表示匹配的授权者名称，path 表示数据路径，#可以代表任何数字，code 表示返回代码。

使用 UriMatcher 时，则可以直接调用 match()函数，对指定的 URI 进行判断。

例如：

```
switch(uriMatcher.match(uri)){
    case MULTIPLE_PEOPLE:
        //多条数据的处理过程
        break;
    case SINGLE_PEOPLE:
```

```
                //单条数据的处理过程
                break;
            default:
                throw new IllegalArgumentException("不支持的 URI:" + uri);
}
```

3. 注册 ContentProvider

SQLiteProvider 直接继承自 ContentProvider，是一个提供数据共享的 ContentProvider 组件，在使用它之前，必须在 AndroidManifest.xml 中注册。

AndroidManifest.xml 代码如下：

```xml
<?xml version="1.0" encoding="utf-8"?>
<manifest xmlns:android="http://schemas.android.com/apk/res/android"
    package="com.easybooks.android.sqlitereg">

    <application
    …
        <activity
            android:name=".RegisterActivity"
        …
        </activity>
        <activity android:name=".SuccessActivity"/>
        <provider
            android:authorities="com.easybooks.android.sqlitereg.SQLiteProvider"
            android:name=".SQLiteProvider"
            android:exported="true"/>
    </application>

</manifest>
```

8.4.3 ContentProvider 应用（【例四】：获取注册信息）

【例四】使用 ContentProvider 共享工程 SqliteReg 的注册用户数据，并另开发一个应用来访问共享的数据。

1. 共享 SQLite 数据

打开工程 SqliteReg，在其中添加一个 SQLiteProvider 提供数据共享。

SQLiteProvider.java 代码如下：

```java
package com.easybooks.android.sqlitereg;

import android.content.ContentProvider;
import android.content.ContentUris;
import android.content.ContentValues;
import android.content.UriMatcher;
import android.database.Cursor;
import android.database.sqlite.SQLiteDatabase;
import android.database.sqlite.SQLiteQueryBuilder;
import android.net.Uri;

public class SQLiteProvider extends ContentProvider {
    //声明一个 SQLiteAdapter 对象作为访问 SQLite 数据库的中介
```

```java
        private SQLiteAdapter sqLiteAdapter;
        private SQLiteDatabase sqLiteDb;                              //数据库实例
        private static final String SQLITE_NAME = "SqliteUserInfo";   //数据库名称
        private static final String TABLE_NAME = "users";             //记录用户信息的表名

        private static final int USERS = 1;
        private static final int USER = 2;
        private static final UriMatcher MATCHER;

        static {
            MATCHER = new UriMatcher(UriMatcher.NO_MATCH);
            MATCHER.addURI("com.easybooks.android.sqlitereg.SQLiteProvider", "users", USERS);
                                                                //不带主键编号的 Uri
            MATCHER.addURI("com.easybooks.android.sqlitereg.SQLiteProvider", "users/#", USER);
                                                                //带主键编号的 Uri
        }

        @Override
        public boolean onCreate() {
            sqLiteAdapter = new SQLiteAdapter(getContext(), SQLITE_NAME);
            sqLiteDb = sqLiteAdapter.getSqliteDb();
            if (sqLiteDb == null) return false;
            else return true;
        }

        /*
         * 返回当前 Uri 所代表数据的 MIME 类型:
         * 如果操作的数据属于集合类型,那么 MIME 类型字符串应该以 vnd.android.cursor.dir/开头
         * 如果操作的数据属于非集合类型,那么 MIME 类型字符串应该以 vnd.android.cursor.item/开头
         */
        @Override
        public String getType(Uri uri) {
            switch (MATCHER.match(uri)) {
                case USERS:
                    return "vnd.android.cursor.dir/vnd.easybooks.users";
                case USER:
                    return "vnd.android.cursor.item/vnd.easybooks.users";
                default:
                    throw new IllegalArgumentException("Failed to getType:" + uri.toString());
            }
        }
        //供外部应用从 ContentProvider 添加数据
        @Override
        public Uri insert(Uri uri, ContentValues values) {
            switch (MATCHER.match(uri)) {
                case USERS:
                    long userId = sqLiteDb.insert(TABLE_NAME, null, values);
                    Uri insertUri = ContentUris.withAppendedId(uri, userId);
                    getContext().getContentResolver().notifyChange(insertUri, null);
                    return insertUri;
                default:
```

```java
            throw new IllegalArgumentException("Failed to insert:" + uri.toString());
    }
}

// 供外部应用从 ContentProvider 删除数据
@Override
public int delete(Uri uri, String selection, String[] selectionArgs) {
    int count = 0;
    switch (MATCHER.match(uri)) {
        case USERS:
            count = sqLiteDb.delete(TABLE_NAME, selection, selectionArgs);
            break;
        case USER:
            String segment = uri.getPathSegments().get(1);
            count = sqLiteDb.delete(TABLE_NAME, "_id=" + segment, selectionArgs);
            break;
        default:
            throw new IllegalArgumentException("Failed to delete:" + uri.toString());
    }
    getContext().getContentResolver().notifyChange(uri, null);
    return count;
}

// 供外部应用更新 ContentProvider 中的数据
@Override
public int update(Uri uri, ContentValues values, String selection, String[] selectionArgs) {
    int count = 0;
    switch (MATCHER.match(uri)) {
        case USERS:
            count = sqLiteDb.update(TABLE_NAME, values, selection, selectionArgs);
            break;
        case USER:
            String segment = uri.getPathSegments().get(1);
            count = sqLiteDb.update(TABLE_NAME, values, "_id=" + segment, selectionArgs);
            break;
        default:
            throw new IllegalArgumentException("Failed to update:" + uri.toString());
    }
    getContext().getContentResolver().notifyChange(uri, null);
    return count;
}

// 供外部应用从 ContentProvider 中获取数据
@Override
public Cursor query(Uri uri, String[] projection, String selection, String[] selectionArgs, String sortOrder) {
    SQLiteQueryBuilder qb = new SQLiteQueryBuilder();
    qb.setTables(TABLE_NAME);
    switch (MATCHER.match(uri)) {
        case USER:
            qb.appendWhere("_id=" + uri.getPathSegments().get(1));
            break;
        default:
            break;
```

```
            }
            Cursor cursor = qb.query(sqLiteDb, projection, selection, selectionArgs, null, null, sortOrder);
            cursor.setNotificationUri(getContext().getContentResolver(), uri);
            return cursor;
        }
    }
}
```

2. 创建登录应用

创建 Android 工程 SqliteLog，直接套用例一开发好的登录应用框架，将相关的源文件及资源复制到本项目对应的目录即可。

只需修改 LoginActivity 的 loadUser 方法并声明对应的全局变量即可。

LoginActivity.java 代码如下（加黑处为修改的部分）：

```java
package com.easybooks.android.sqlitelog;

import android.content.ContentResolver;
…

public class LoginActivity extends AppCompatActivity {
    private EditText myName;
    private EditText myPwd;
    private Button myOk;
    private TimePicker myTime;
    private DatePicker myDate;
    private static final String CONTENT_URI = "content://com.easybooks.android.sqlitereg.SQLiteProvider/users/1";
    private ContentResolver resolver = null;
    private String myusername;//用户名
    private String mypassword;//密码

    @Override
    protected void onCreate(Bundle savedInstanceState) {
        super.onCreate(savedInstanceState);
        setContentView(R.layout.login);
        findViews();
        loadUser();
    }

    private void findViews() {          …      }

    private void loadUser() {
        resolver = getContentResolver();
        Uri uri = Uri.parse(CONTENT_URI);
        Cursor cursor = resolver.query(uri, new String[]{"name", "pwd"}, null, null, null);
        if (cursor != null) {
            cursor.moveToFirst();
            myusername = cursor.getString(0);
            myName.setText(myusername);
            mypassword = cursor.getString(cursor.getColumnIndex("pwd"));
        }
    }
```

```
public void onLoginClick(View view) {
    …
    }
}
```

3. 运行效果

运行效果同例一的登录应用，见图 8.3 和图 8.4。

第9章 Android 地图应用开发

在 Android 系统下,地图是移动终端的主要应用。例如,根据你当前位置给你提供最近的各种服务商店,到你指定位置的最佳路径等。

目前,比较流行的地图服务主要包括百度、高德、谷歌等。本书主要先以百度地图进行介绍,在此基础上,简要介绍高德地图。

9.1 创建地图开发环境

9.1.1 百度地图环境

1. 下载地图 SDK

访问百度地图开放平台的 SDK 下载页 http://lbsyun.baidu.com/index.php?title=androidsdk/sdkandev-download,如图 9.1 所示,点"一键下载"。

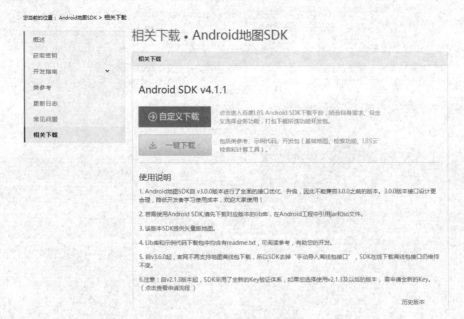

图 9.1 百度地图 SDK 下载页

下载得到名为 BaiduMap_AndroidSDK_v4.1.1_All 的压缩包,解压进入 libs 目录,可见如图 9.2 所示的目录结构。

其中包含 6 个.jar 包和 5 个文件夹,.jar 包是开发地图必不可少的类库,文件夹中则是对应各手机平台的.so 库。

第 9 章 Android 地图应用开发

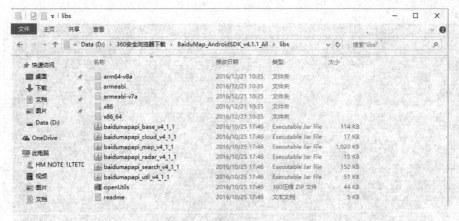

图 9.2　SDK 的目录结构

2. 将 SDK 加载到项目中

创建 Android 项目，名称为 MyBd，切换到 Project 模式，如图 9.3 所示，将图 9.2 目录下的 6 个 .jar 包复制到项目 app\libs 下。在 app\src\main 下新建一个名为"jniLibs"的文件夹，将图 9.2 目录下的 5 个文件夹全部复制进去。最后选择 app\libs 下的所有 .jar 包，按右键，选择 add as library，按"OK"。

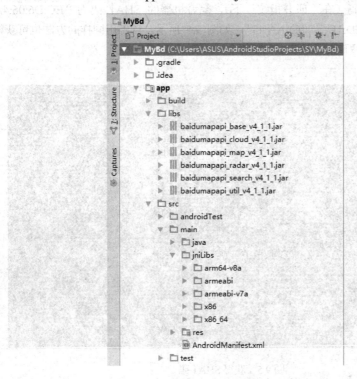

图 9.3　将 SDK 加载进项目

3. 获取本地 SHA1

打开 Windows 命令行，输入 cd .android 进入本地计算机的 Android SDK 安装目录。

输入 keytool -v -list -keystore debug.keystore 回车，出现"输入密钥库口令"提示，直接回车，如图 9.4 所示。

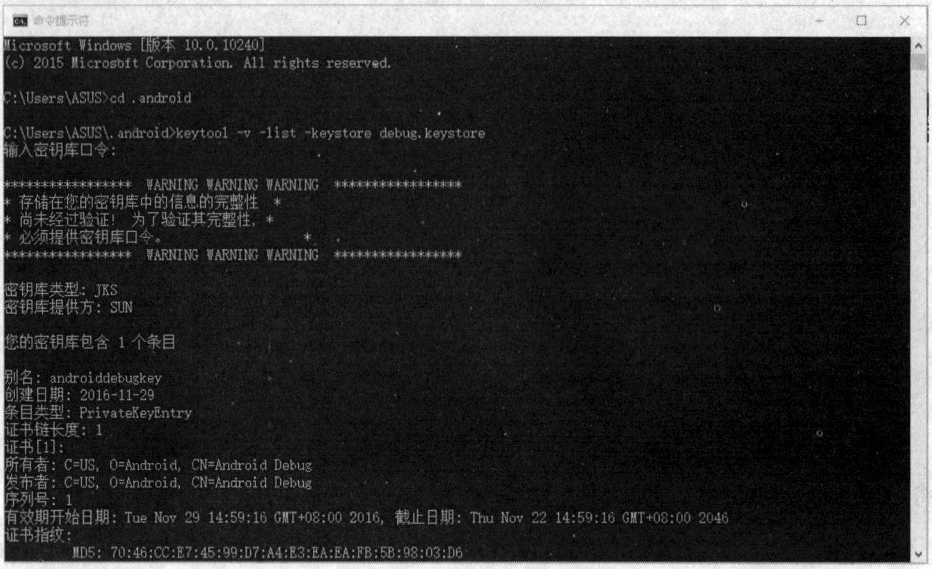

图 9.4 访问本机密钥库

于是可以获得 SHA1 码（在"证书指纹"下），笔者机器的 SHA1 码为"BC:D6:06:4A:F2:2D: AE:0A:4B:DB:92:04:75:A6:ED:1D:EC:0D:AA:17"，如图 9.5 所示，读者用同样的方法也可获得自己计算机的 SHA1，请务必将它复制记录下来，以备后用。

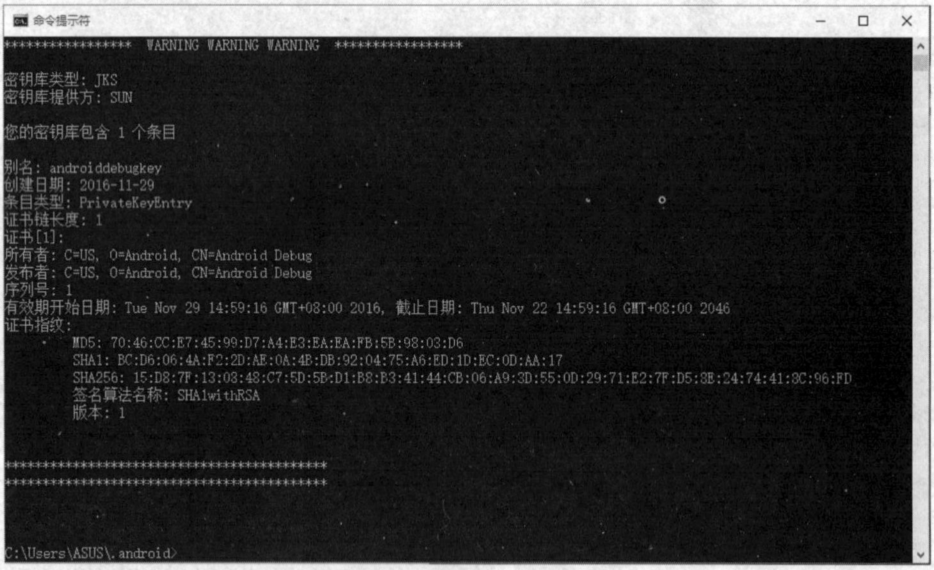

图 9.5 获得 SHA1 码

4. 登录百度账号创建应用

访问百度地图开放平台官网：http://lbsyun.baidu.com/，用自己的百度账号登录（读者若没有百度账号，请马上注册一个），创建应用，可看到如图 9.6 所示的页面。

第 9 章　Android 地图应用开发

图 9.6　创建应用

其中：
（1）应用名称：创建的项目名 MyBd。
（2）应用类型：选"Android SDK"。
（3）发布版 SHA1 和开发版 SHA1：均填写刚刚获取记录的本地 SHA1。
（4）包名：就是项目 MyBd 自身的包名，可在项目 AndroidManifest.xml 文件中查看到，如图 9.7 所示，这里是"com.easybooks.android.mybd"。

图 9.7　查看项目所在的包名

填写完以上各项信息后，单击"提交"，系统生成应用并自动产生了该应用的访问 API Key（简称 AK），如图 9.8 所示。

图 9.8　创建成功生成 AK

可以看到，本应用的访问 AK 值为"kcdxsS7otjNYu9iZmcbaHAki0PbamSP2"（读者请以自己实际操作所生成的 AK 值为准），可用记事本记录下来以备后用。

5. 配置工程环境

（1）打开工程 MyBd，在 Android Studio 主菜单栏选择 File→Project Structure，在弹出的窗口中配置项目的 keystore，如图 9.9 所示。

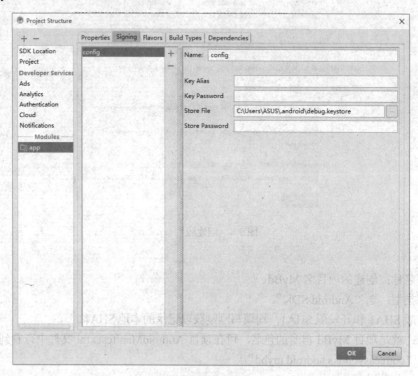

图 9.9　配置项目的 keystore

（2）在 AndroidManifest.xml 中添加开发密钥、所需权限等信息。

AndroidManifest.xml 代码如下（加黑处为需要添加配置的内容）：

```
<?xml version="1.0" encoding="utf-8"?>
<manifest xmlns:android="http://schemas.android.com/apk/res/android"
    package="com.easybooks.android.mybd">
    <uses-permission android:name="android.permission.ACCESS_NETWORK_STATE"/>
    <uses-permission android:name="android.permission.INTERNET"/>
    <uses-permission android:name="android.permission.READ_SYNC_SETTINGS"/>
    <uses-permission android:name="android.permission.WAKE_LOCK"/>
    <uses-permission android:name="android.permission.CHANGE_WIFI_STATE"/>
    <uses-permission android:name="android.permission.ACCESS_WIFI_STATE"/>
    <uses-permission android:name="android.permission.GET_TASKS"/>
    <uses-permission android:name="android.permission.WRITE_EXTERNAL_STORAGE"/>
    <uses-permission android:name="android.permission.WRITE_SETTINGS"/>
    <uses-permission android:name="android.permission.READ_PHONE_STATE"/>
    <application
        android:allowBackup="true"
        android:icon="@mipmap/ic_launcher"
        android:label="@string/app_name"
```

```xml
            android:supportsRtl="true"
            android:theme="@style/AppTheme">
            <meta-data android:name="com.baidu.lbsapi.API_KEY"
                android:value="kcdxsS7otjNYu9iZmcbaHAki0PbamSP2"/>
        <activity
            android:name=".MainActivity"
            android:label="@string/app_name"
            android:theme="@style/AppTheme.NoActionBar">
            <intent-filter>
                <action android:name="android.intent.action.MAIN" />
                <category android:name="android.intent.category.LAUNCHER" />
            </intent-filter>
        </activity>
    </application>
</manifest>
```

其中，<meta-data>元素的"android:value"属性值即为刚刚在百度地图开放平台创建应用时系统所产生的 AK 值，读者请配置自己实际获得的 AK 值。

6. 使用百度地图功能

经过上述一系列的步骤，终于可以在自己开发的项目中使用百度地图功能了。下面实现最简单的地图显示。

（1）在布局 xml 文件中添加地图控件

修改项目的 content_main.xml 文件源码，代码如下：

```xml
<?xml version="1.0" encoding="utf-8"?>
<RelativeLayout xmlns:android="http://schemas.android.com/apk/res/android"
    …>
    <com.baidu.mapapi.map.MapView
        android:id="@+id/bmapView"
        android:layout_width="fill_parent"
        android:layout_height="fill_parent"
        android:clickable="true"/>
</RelativeLayout>
```

其中，"<com.baidu.mapapi.map.MapView/>"为百度地图 API 提供的地图控件，Android Studio 环境本身没有，需要用户自己编写 XML 添加该元素。

（2）在 Activity 中创建地图，管理地图生命周期

修改 MainActivity.java，代码如下：

```java
package com.easybooks.android.mybd;
import android.os.Bundle;
…
import com.baidu.mapapi.SDKInitializer;
import com.baidu.mapapi.map.MapView;
public class MainActivity extends AppCompatActivity {
    private MapView myMap = null;              //声明地图控件
    @Override
    protected void onCreate(Bundle savedInstanceState) {
        super.onCreate(savedInstanceState);
        //在使用 SDK 各组件之前初始化 context 信息，传入 ApplicationContext
        //注意该方法要在 setContentView 方法之前实现
        SDKInitializer.initialize(getApplicationContext());
```

```
        setContentView(R.layout.activity_main);
        //获取地图控件引用
        myMap = (MapView) findViewById(R.id.bmapView);
        Toolbar toolbar = (Toolbar) findViewById(R.id.toolbar);
        …
    }
    @Override
    protected void onResume() {
        super.onResume();
        //在 activity 执行 onResume 时执行 myMap.onResume(),实现地图生命周期管理
        myMap.onResume();
    }
    @Override
    protected void onPause() {
        super.onPause();
        //在 activity 执行 onPause 时执行 myMap.onPause(),实现地图生命周期管理
        myMap.onPause();
    }
    @Override
    protected void onDestroy() {
        super.onDestroy();
        //在 activity 执行 onDestroy 时执行 myMap.onDestroy(),实现地图生命周期管理
        myMap.onDestroy();
    }
    @Override
    public boolean onCreateOptionsMenu(Menu menu) {  …  }
    @Override
    public boolean onOptionsItemSelected(MenuItem item) {  …  }
}
```

运行程序,显示地图,如图 9.10 所示。

图 9.10 显示百度地图

9.1.2 高德地图环境

创建高德地图开发环境的步骤与百度地图大同小异,为了方便读者,现也详细介绍如下。

1. 下载地图 SDK

访问高德开放平台的 SDK 下载页 http://lbs.amap.com/api/android-sdk/download/,如图 9.11 所示,点"Android 地图 SDK 一键下载"。

下载得到名为"AMap_Android_SDK_All"的压缩包,解压看到其中又有 6 个压缩包,如图 9.12 所示。

其中,AMap2Dmap、AMap3Dmap 和 AmapSearch 这 3 个包分别解压,得到文件夹中的.jar 包是开发地图必不可少的类库,而 AMap3Dmap 解压目录下的两个子文件夹中则是.so 库。

另外三个包是 SDK 自带的文档和范例,并非构建地图开发环境所必需。读者也可解压后用作开发时的学习参考资料。

第9章 Android地图应用开发

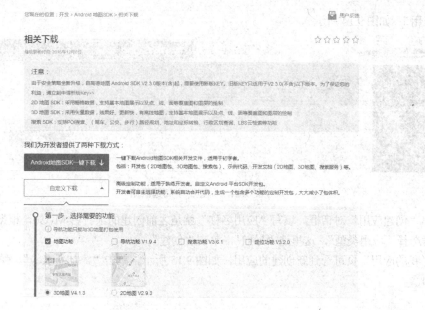

图9.11 高德地图SDK下载页

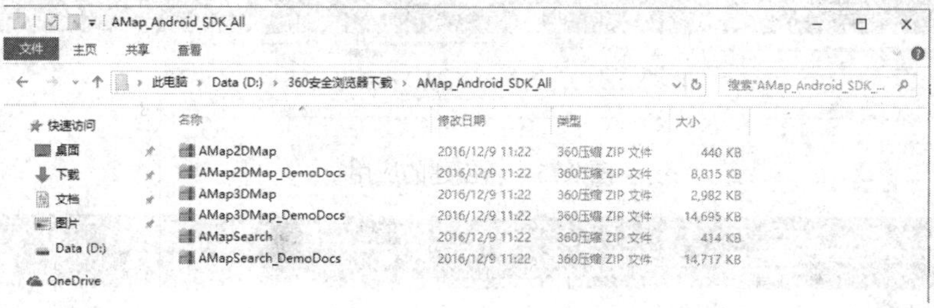

图9.12 SDK中的压缩包

2. 将SDK加载到项目中

创建Android项目，名称为"MyGd"，切换到Project模式，如图9.13所示，将上一步解压各压缩包得到的.jar包（一共3个）复制到项目app\libs下。在app\src\main下新建一个名为"jniLibs"的文件夹，将AMap3Dmap解压目录下的两个子文件夹复制进去。最后选择app\libs下所有.jar包，按右键，选择add as library，按"OK"即可。

3. 获取本地SHA1

操作过程同9.1.1百度地图开发的第3步，本地计算机的SHA1是唯一的，如果用户在之前开发中已进行过这一步，可将SHA1码记录在案，之后从事地图开发都使用同一个SHA1码，无须重复获取过程。

4. 登录高德账号创建应用

访问高德开放平台官网：http://lbs.amap.com/，用自己的高德账号登录（读者若没有高德账号，请马上注册一个），点"新

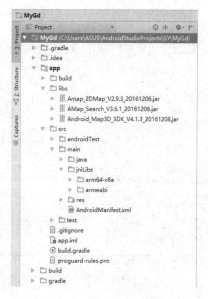

图9.13 将SDK加载进项目

建一个应用",如图 9.14 所示。

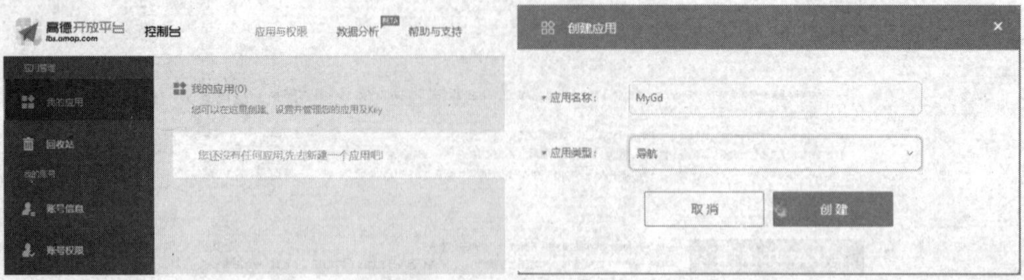

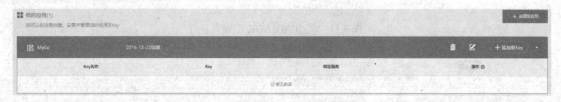

图 9.14 新建一个高德应用

弹出"创建应用"对话框,填写"应用名称"就是之前创建的项目名 MyGd,根据所开发程序的实际用途选择"应用类型",这里选"导航",单击"创建"。

在"我的应用"页可看到新创建的应用,如图 9.15 所示。单击"添加新 Key",弹出如图 9.16 所示的窗口。

图 9.15 新创建的应用

图 9.16 为应用添加 Key

其中：

（1）Key 名称：填写 MyGdKey。

（2）服务平台：选"Android 平台 SDK"。

（3）发布版安全码 SHA1 和调试版安全码 SHA1：均填写之前获取的本地 SHA1。

（4）Package：就是项目 MyGd 自身的包名，同样可在当前工程 AndroidManifest.xml 文件中查看，如图 9.17 所示，这里是"com.easybooks.android.mygd"。

图 9.17　查看项目所在的包名

填写完以上各项信息后，勾选"我已阅读高德地图 API 服务条款"，单击"提交"，系统生成应用并自动产生了该应用的访问 Key，如图 9.18 所示。

图 9.18　创建成功生成 Key

可以看到，本应用的访问 Key 值为"b100754c1fd2cef881371ee370b842ec"（读者请以自己实际操作所生成的 Key 值为准），可用记事本记录下来以备后用。

5. 配置工程环境

（1）打开工程 MyGd，在 Android Studio 主菜单栏选择 File→Project Structure，在弹出的窗口中配置项目的 keystore，同图 9.9，此处略。

（2）在 AndroidManifest.xml 中添加开发密钥、所需权限等信息。

AndroidManifest.xml 代码如下（加黑处为需要添加配置的内容）：

```
<?xml version="1.0" encoding="utf-8"?>
<manifest xmlns:android="http://schemas.android.com/apk/res/android"
    package="com.easybooks.android.mygd">
    <uses-permission android:name="android.permission.INTERNET"/>
    <uses-permission android:name="android.permission.WRITE_EXTERNAL_STORAGE"/>
    <uses-permission android:name="android.permission.ACCESS_NETWORK_STATE"/>
    <uses-permission android:name="android.permission.ACCESS_WIFI_STATE"/>
    <uses-permission android:name="android.permission.READ_PHONE_STATE"/>
    <uses-permission android:name="android.permission.ACCESS_COARSE_LOCATION"/>
    <application
        …
        android:theme="@style/AppTheme">
        <meta-data android:name="com.amap.api.v2.apikey"
```

```xml
            android:value="b100754c1fd2cef881371ee370b842ec"/>
        <activity
            android:name=".MainActivity"
            …
        </activity>
    </application>
</manifest>
```

其中，<meta-data>元素的"android:value"属性值即为刚刚在高德开放平台创建应用时系统产生的 Key 值，读者请配置自己实际获得的 Key 值。

6. 使用高德地图功能

经过上述一系列的步骤，可以在自己开发的项目中使用高德地图功能了。下面实现最简单的地图显示。

(1) 在布局 xml 文件中添加地图控件

修改项目的 content_main.xml 文件源码，代码如下：

```xml
<?xml version="1.0" encoding="utf-8"?>
<RelativeLayout xmlns:android="http://schemas.android.com/apk/res/android"
    …>
    <com.amap.api.maps.MapView
        android:id="@+id/gmapView"
        android:layout_width="fill_parent"
        android:layout_height="fill_parent"
        android:clickable="true" />
</RelativeLayout>
```

(2) 在 Activity 中创建地图，管理地图生命周期

修改 MainActivity.java，代码如下：

```java
package com.easybooks.android.mygd;
…
import com.amap.api.maps.MapView;
public class MainActivity extends AppCompatActivity {
    private MapView myMap = null;                        //声明地图控件
    @Override
    protected void onCreate(Bundle savedInstanceState) {
        super.onCreate(savedInstanceState);
        setContentView(R.layout.activity_main);
        myMap = (MapView) findViewById(R.id.gmapView);    //获取地图控件引用
        myMap.onCreate(savedInstanceState);              //此方法必须重写
        Toolbar toolbar = (Toolbar) findViewById(R.id.toolbar);
        …
    }
    @Override
    protected void onResume() {
        super.onResume();
        //在 activity 执行 onResume 时执行 myMap.onResume()，实现地图生命周期管理
        myMap.onResume();
    }
    @Override
    protected void onPause() {
        super.onPause();
```

```
            //在 activity 执行 onPause 时执行 myMap.onPause()，实现地图生
命周期管理
            myMap.onPause();
        }
        @Override
        protected void onDestroy() {
            super.onDestroy();
            //在 activity 执行 onDestroy 时执行 myMap.onDestroy()，实现地图
生命周期管理
            myMap.onDestroy();
        }
            …
    }
```

运行程序，显示地图，如图 9.19 所示。

为了避免反复注册应用和申请 Key 值的麻烦，读者请将已经创建的地图 Android 项目各复制一份，后面的实例将在其基础上做。

本章着重以百度地图为例介绍地图各种常用功能的开发，故下面的每个实例都在之前创建的 MyBd 项目的基础上做。

图 9.19　显示高德地图

9.2　设置地图类型及区域检索

【例 9.1】根据选择的城市和区域检索，按照地图类型显示。

9.2.1　设计界面

界面设计视图如图 9.20 所示。

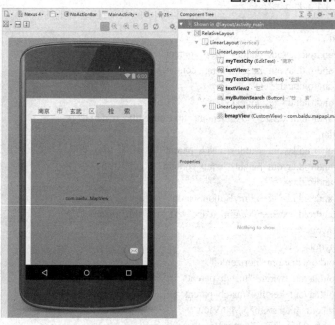

图 9.20　区域检索设计视图

它所对应的 content_main.xml 的代码如下：

```xml
<?xml version="1.0" encoding="utf-8"?>
<RelativeLayout xmlns:android="http://schemas.android.com/apk/res/android"
    …>
    <LinearLayout
        android:orientation="vertical"
        android:layout_width="match_parent"
        android:layout_height="match_parent">
        <LinearLayout
            android:orientation="horizontal"
            android:layout_width="match_parent"
            android:layout_height="wrap_content"
            android:gravity="center_horizontal">
            <EditText
                android:layout_width="80dp"
                android:layout_height="wrap_content"
                android:id="@+id/myTextCity"
                android:text="南京"
                android:textAlignment="center" />
            <TextView
                android:layout_width="wrap_content"
                android:layout_height="wrap_content"
                android:text="市"
                android:id="@+id/textView"
                android:textSize="@dimen/abc_text_size_title_material" />
            <EditText
                android:layout_width="80dp"
                android:layout_height="wrap_content"
                android:id="@+id/myTextDistrict"
                android:text="玄武"
                android:textAlignment="center" />
            <TextView
                android:layout_width="wrap_content"
                android:layout_height="wrap_content"
                android:text="区"
                android:id="@+id/textView2"
                android:textSize="@dimen/abc_text_size_title_material" />
            <Button
                android:layout_width="match_parent"
                android:layout_height="wrap_content"
                android:text="检    索"
                android:id="@+id/myButtonSearch"
                android:textSize="@dimen/abc_text_size_title_material" />
        </LinearLayout>
        <LinearLayout
            android:orientation="horizontal"
            android:layout_width="match_parent"
            android:layout_height="match_parent">
            <com.baidu.mapapi.map.MapView
                android:id="@+id/bmapView"
                android:layout_width="fill_parent"
                android:layout_height="fill_parent"
```

```
                android:clickable="true"/>
        </LinearLayout>
    </LinearLayout>
</RelativeLayout>
```

界面很简洁，主要使用的是线性布局及本书之前章节介绍的 EditText、TextView、Button 等 Android 开发最常用的控件，只不过在最后添加了一个用于显示地图的 MapView 控件，在此就不做过多的介绍了。

为了能让用户设置地图的各种不同显示类型及自定义百度地图 Logo 的显示位置，本应用设计了一个菜单系统，其 menu_main.xml 源码如下：

```xml
<menu xmlns:android="http://schemas.android.com/apk/res/android"
    xmlns:app="http://schemas.android.com/apk/res-auto"
    xmlns:tools="http://schemas.android.com/tools"
    tools:context="com.easybooks.android.mybd.MainActivity">
    <item
        android:id="@+id/main_menu_0"
        android:orderInCategory="0"
        android:title="普通模式" />
    <item
        android:id="@+id/main_menu_1"
        android:orderInCategory="1"
        android:title="卫星模式" />
    <item
        android:id="@+id/main_menu_2"
        android:orderInCategory="2"
        android:title="实时路况" />
    <item
        android:id="@+id/main_menu_3"
        android:orderInCategory="3"
        android:title="城市热力" />
    <item
        android:id="@+id/main_menu_4"
        android:orderInCategory="4"
        android:title="百度 Logo">
        <menu>
            <item
                android:id="@+id/sub_menu_4_0"
                android:title="右上" />
            <item
                android:id="@+id/sub_menu_4_1"
                android:title="左下" />
            <item
                android:id="@+id/sub_menu_4_2"
                android:title="中上" />
            <item
                android:id="@+id/sub_menu_4_3"
                android:title="中下" />
            <item
                android:id="@+id/sub_menu_4_4"
                android:title="左上" />
            <item
                android:id="@+id/sub_menu_4_5"
```

```xml
                    android:title="右下" />
                </menu>
            </item>
            <item
                android:id="@+id/main_menu_5"
                android:orderInCategory="5"
                android:title="所有标注" />
</menu>
```

选择不同的菜单项，分别可切换显示普通/卫星地图、实时路况图、城市热力图，图中百度 Logo 的位置也可根据实际需要置于右上、左下、中上、中下、左上、右下等 6 个不同的位置，还可选择去除地图上的所有标注信息。

9.2.2 功能实现

地图功能实现代码全部位于 MainActivity.java 源文件中。

当输入城市名和检索区域后，单击"检索"按钮，在其事件代码中执行 DistrictSearch 对象的 searchDistrict()方法（加黑语句），其中传入城市名和区域名称两个参数。然后在 onGetDistrictResult() 方法中处理返回的检索结果。

而对地图类型的设置则通过重写系统菜单方法 onOptionsItemSelected()，在 switch 语句中针对用户选择的不同菜单项显示不同的地图类型。

MainActivity.java 代码如下：

```java
package com.easybooks.android.mybd;
import …
public class MainActivity extends AppCompatActivity {
    private MapView myMap = null;
    private BaiduMap myBaiduMap = null;
    private EditText myCity;
    private EditText myDistrict;
    private Button mySearch;
    private boolean trafficOn = false;           //开关实时路况显示
    private boolean heatMapOn = false;           //开关城市热力（人口密度）显示
    private boolean showPoiOn = true;            //地图上显示（清除）所有标注
    private DistrictSearch myDistrictSearch;     //(1)
    @Override
    protected void onCreate(Bundle savedInstanceState) {
        super.onCreate(savedInstanceState);
        //在使用 SDK 各组件之前初始化 context 信息，传入 ApplicationContext
        //注意该方法要在 setContentView 方法之前实现
        SDKInitializer.initialize(getApplicationContext());
        setContentView(R.layout.activity_main);
        //获取地图控件引用
        myMap = (MapView) findViewById(R.id.bmapView);
        myBaiduMap = myMap.getMap();
        //初始化
        findViews();
        myDistrictSearch = DistrictSearch.newInstance();        //(2)
        Toolbar toolbar = (Toolbar) findViewById(R.id.toolbar);
        …
        //单击"检索"按钮的事件方法代码
```

```java
        mySearch.setOnClickListener(new View.OnClickListener() {
            @Override
            public void onClick(View v) {
                String city = "";
                String district = "";
                if (myCity.getText() != null && !"".equals(myCity.getText())) {
                    city = myCity.getText().toString();        //获取城市名
                }
                if (myDistrict.getText() != null && !"".equals(myDistrict.getText())) {
                    district = myDistrict.getText().toString();   //获取城区名
                }
                myDistrictSearch.searchDistrict(new    DistrictSearchOption().cityName(city).districtName(district));                                    //(3)
            }
        });
        //在回调方法中获取并处理区域检索的结果
        myDistrictSearch.setOnDistrictSearchListener(new OnGetDistricSearchResultListener() {
            @Override
            public void onGetDistrictResult(DistrictResult districtResult) {
                myBaiduMap.clear();
                if (districtResult == null) {
                    return;
                }
                if (districtResult.error == SearchResult.ERRORNO.NO_ERROR) {         //(4)
                    List<List<LatLng>> polyLines = districtResult.getPolylines();    //(5)
                    if (polyLines == null) {
                        return;
                    }
                    LatLngBounds.Builder builder = new LatLngBounds.Builder();
                    for (List<LatLng> polyline : polyLines) {                         //(6)
                        OverlayOptions ooPolyline11 = new PolylineOptions().width(10).points(polyline).dottedLine(true).color(0xAA00FF00);
                        myBaiduMap.addOverlay(ooPolyline11);                          //(7)
                        OverlayOptions ooPolygon = new PolygonOptions().points(polyline).stroke(new Stroke(5, 0xAA00FF88)).fillColor(0xAAFFFF00);       //(8)
                        myBaiduMap.addOverlay(ooPolygon);
                        for (LatLng latLng : polyline) {
                            builder.include(latLng);
                        }
                    }
                    myBaiduMap.setMapStatus(MapStatusUpdateFactory.newLatLngBounds(builder.build()));
                }
            }
        });
    }
    private void findViews() {
        myCity = (EditText) findViewById(R.id.myTextCity);
        myDistrict = (EditText) findViewById(R.id.myTextDistrict);
        mySearch = (Button) findViewById(R.id.myButtonSearch);
    }
    @Override
    protected void onResume() { ... }
```

```java
        @Override
        protected void onPause() { … }
        @Override
        protected void onDestroy() {
            myDistrictSearch.destroy();                                              //(9)
            super.onDestroy();
            //在 activity 执行 onDestroy 时执行 myMap.onDestroy()，实现地图生命周期管理
            myMap.onDestroy();
        }
        @Override
        public boolean onCreateOptionsMenu(Menu menu) {   …   }
        @Override
        public boolean onOptionsItemSelected(MenuItem item) {      //创建菜单
            switch (item.getItemId()) {
                case R.id.main_menu_0:                             //普通模式
                    myBaiduMap.setMapType(BaiduMap.MAP_TYPE_NORMAL);
                    return true;
                case R.id.main_menu_1:                             //卫星模式
                    myBaiduMap.setMapType(BaiduMap.MAP_TYPE_SATELLITE);
                    return true;
                case R.id.main_menu_2:                             //实时路况
                    if (!trafficOn)
                        myBaiduMap.setTrafficEnabled(true);
                    else myBaiduMap.setTrafficEnabled(false);
                    trafficOn = !trafficOn;
                    return true;
                case R.id.main_menu_3:                             //城市热力
                    if (!heatMapOn)
                        myBaiduMap.setBaiduHeatMapEnabled(true);
                    else myBaiduMap.setBaiduHeatMapEnabled(false);
                    heatMapOn = !heatMapOn;
                    return true;
                case R.id.sub_menu_4_0:                            //右上
                    myMap.setLogoPosition(LogoPosition.logoPostionRightTop);
                    return true;
                case R.id.sub_menu_4_1:                            //左下
                    myMap.setLogoPosition(LogoPosition.logoPostionleftBottom);
                    return true;
                case R.id.sub_menu_4_2:                            //中上
                    myMap.setLogoPosition(LogoPosition.logoPostionCenterTop);
                    return true;
                case R.id.sub_menu_4_3:                            //中下
                    myMap.setLogoPosition(LogoPosition.logoPostionCenterBottom);
                    return true;
                case R.id.sub_menu_4_4:                            //左上
                    myMap.setLogoPosition(LogoPosition.logoPostionleftTop);
                    return true;
                case R.id.sub_menu_4_5:                            //右下
                    myMap.setLogoPosition(LogoPosition.logoPostionRightBottom);
                    return true;
                case R.id.main_menu_5:                             //所有标注
                    if (!showPoiOn)
```

```
                myBaiduMap.showMapPoi(true);
            else myBaiduMap.showMapPoi(false);
            showPoiOn = !showPoiOn;
            return true;
        default:
            return false;
        }
    }
}
```

其中：

（1）百度地图的区域检索使用 DistrictSearch 对象，需要事先声明。

（2）DistrictSearch 对象在使用前要先实例化。

（3）调用 DistrictSearch 对象的 searchDistrict()方法来获取城市及区域信息。

（4）区域检索的结果在 DistrictResult 类型的引用参数中返回，若返回结果为 SearchResult.ERRORNO.NO_ERROR 表示正确无误。

（5）DistrictResult 类的 getPolylines()方法获取检索区域边界线上全部点的经纬度。

（6）以这些点的经纬度值为依据，描绘出区域边界曲线。

（7）将边界线作为覆盖物添加到地图上。

（8）给该区域填充色彩，并且也以覆盖物的形式添加到地图上。

（9）DistrictSearch 对象在程序结束时要及时销毁，结束其生命周期。

9.2.3 运行效果

运行程序，输入"南京"市"玄武"区，点"检索"按钮，程序以淡黄色覆盖标识出南京市玄武区的轮廓，如图 9.21 所示。

点击标题栏右上角，出现如图 9.22 所示的菜单，点选其中各个菜单项，分别可切换到卫星模式（见图 9.23）、实时路况（见图 9.24）、城市热力（见图 9.25）以及清除地图上的所有标注信息（见图 9.26）。

图 9.21 区域检索　　　　　　图 9.22 功能菜单　　　　　　图 9.23 卫星模式

图 9.24　实时路况　　　　　图 9.25　城市热力　　　　　图 9.26　去除所有标注信息

9.3　地理经纬度检索

【例 9.2】根据经纬度查找地址，或者按照地址查找经纬度。

9.3.1　设计界面

设计视图如图 9.27 所示。

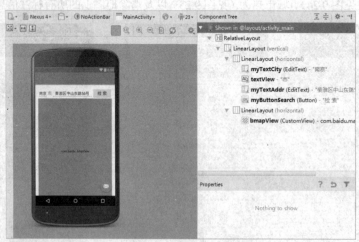

图 9.27　经纬度检索设计视图

content_main.xml 源码如下：

```
<?xml version="1.0" encoding="utf-8"?>
<RelativeLayout xmlns:android="http://schemas.android.com/apk/res/android"
       ...>
    <LinearLayout
```

```xml
            android:orientation="vertical"
            android:layout_width="match_parent"
            android:layout_height="match_parent">
        <LinearLayout
            android:orientation="horizontal"
            android:layout_width="match_parent"
            android:layout_height="wrap_content">
            <EditText
                android:layout_width="45dp"
                android:layout_height="wrap_content"
                android:id="@+id/myTextCity"
                android:text="南京"
                android:textAlignment="center" />
            <TextView
                android:layout_width="wrap_content"
                android:layout_height="wrap_content"
                android:text="市"
                android:id="@+id/textView"
                android:textSize="@dimen/abc_text_size_title_material" />
            <EditText
                android:layout_width="195dp"
                android:layout_height="wrap_content"
                android:id="@+id/myTextAddr"
                android:text="秦淮区中山东路56号"
                android:textAlignment="center" />
            <Button
                android:layout_width="match_parent"
                android:layout_height="wrap_content"
                android:text="检 索"
                android:id="@+id/myButtonSearch"
                android:textSize="@dimen/abc_text_size_title_material" />
        </LinearLayout>
        <LinearLayout
            android:orientation="horizontal"
            android:layout_width="match_parent"
            android:layout_height="match_parent">
            <com.baidu.mapapi.map.MapView
                android:id="@+id/bmapView"
                android:layout_width="fill_parent"
                android:layout_height="fill_parent"
                android:clickable="true"/>
        </LinearLayout>
    </LinearLayout>
</RelativeLayout>
```

将 icon_marka.png 图片（见图 9.28）预先复制到项目 app\src\main\res\drawable 目录下。

图 9.28　需要准备的图片资源

9.3.2 功能实现

单击"检索"按钮时,如果用户输入的是地址(不含","),则执行 GeoCoder 对象的 geocode() 方法进行经纬度检索。否则(含","表示用户输入的是经纬度值)用 reverseGeoCode() 方法反向查询出地址信息。

onGetGeoCodeResult() 处理检索到的经纬度数据;onGetReverseGeoCodeResult() 则处理经纬度反向检索到的中文地址信息。两者均通过 addText() 方法显示在界面上。

MainActivity.java 代码如下:

```java
package com.easybooks.android.mybd;
import ...
public class MainActivity extends AppCompatActivity {
    private MapView myMap = null;
    private BaiduMap myBaiduMap = null;
    private EditText myCity;
    private EditText myAddr;
    private Button mySearch;
    private GeoCoder myGeoCoderSearch;         //(1)
    private String lat, lon;                   //经纬度
    @Override
    protected void onCreate(Bundle savedInstanceState) {
        super.onCreate(savedInstanceState);
        //在使用 SDK 各组件之前初始化 context 信息,传入 ApplicationContext
        //注意该方法要在 setContentView 方法之前实现
        SDKInitializer.initialize(getApplicationContext());
        setContentView(R.layout.activity_main);
        //获取地图控件引用
        myMap = (MapView) findViewById(R.id.bmapView);
        myBaiduMap = myMap.getMap();
        //初始化
        findViews();
        myGeoCoderSearch = GeoCoder.newInstance();
        Toolbar toolbar = (Toolbar) findViewById(R.id.toolbar);
        ...
        //单击"检索"按钮的事件方法代码
        mySearch.setOnClickListener(new View.OnClickListener() {
            @Override
            public void onClick(View v) {
                String addr = myAddr.getText().toString();
                if (!addr.contains(",")) {
                    //Geo 搜索(地址名→经纬度)
                    myGeoCoderSearch.geocode(new GeoCodeOption().city(myCity.getText().toString().
address(addr));                                                                   //(2)
                } else {
                    //反 Geo 搜索(经纬度→地址名)
                    String[] laton = addr.split(",");
                    lat = laton[0];
                    lon = laton[1];
                    LatLng ptCenter = new LatLng((Float.valueOf(lat)), (Float.valueOf(lon)));
                    myGeoCoderSearch.reverseGeoCode(new ReverseGeoCodeOption().location(ptCenter));
```

```java
                    }
                }
            });
            //在回调方法中获取并处理地理编码检索的结果
            myGeoCoderSearch.setOnGetGeoCodeResultListener(new OnGetGeoCoderResultListener() {
                @Override
                public void onGetGeoCodeResult(GeoCodeResult geoCodeResult) {        //(4)
                    if (geoCodeResult == null || geoCodeResult.error != SearchResult.ERRORNO.NO_ERROR)
                    {
                        Toast.makeText(MainActivity.this, "未能找到结果", Toast.LENGTH_LONG).show();
                        return;
                    }
                    myBaiduMap.clear();
                    myBaiduMap.addOverlay(new MarkerOptions().position(geoCodeResult.getLocation()).icon
(BitmapDescriptorFactory.fromResource(R.drawable.icon_marka)));                     //(5)
                    myBaiduMap.setMapStatus(MapStatusUpdateFactory.newLatLng(geoCodeResult.getLocation()));
                    //获得经纬度
                    lat = String.format("%.6f", geoCodeResult.getLocation().latitude);
                                                                                    //纬度
                    lon = String.format("%.6f", geoCodeResult.getLocation().longitude);
                                                                                    //经度
                    //将经纬度值以文字覆盖的形式标注在地图上对应的位置
                    String text = "纬度:" + lat + ",经度:" + lon;
                    addText(text);
                }

                @Override
                public void onGetReverseGeoCodeResult(ReverseGeoCodeResult reverseGeoCodeResult) {
                    if (reverseGeoCodeResult == null || reverseGeoCodeResult.error != SearchResult.ERRORNO.NO_ERROR) {
                        Toast.makeText(MainActivity.this, "未能找到结果", Toast.LENGTH_LONG).show();
                        return;
                    }
                    myBaiduMap.clear();
                    myBaiduMap.addOverlay(new MarkerOptions().position(reverseGeoCodeResult.getLocation()).icon(BitmapDescriptorFactory.fromResource(R.drawable.icon_marka)));
                    myBaiduMap.setMapStatus(MapStatusUpdateFactory.newLatLng(reverseGeoCodeResult.getLocation()));
                    //将该地点的名称以文字覆盖的形式标注在地图上对应的位置
                    addText(reverseGeoCodeResult.getAddress());
                }
            });
        }
        private void findViews() {   ...   }

        private void addText(String text) {
            //定义文字所在的坐标点
            LatLng latonText = new LatLng(Float.valueOf(lat), Float.valueOf(lon));
            //构建文字 Option 对象，用于在地图上添加文字
            OverlayOptions textOption = new TextOptions().bgColor(0xAAFFFF00).fontSize(24).fontColor
```

```
(0xFFFF00FF).text(text).rotate(-30).position(latonText);
        //在地图上添加该文字对象并显示
        myBaiduMap.addOverlay(textOption);
    }
    …
    @Override
    protected void onDestroy() {
        myGeoCoderSearch.destroy();
        super.onDestroy();
    …
    }
    …
}
```

其中：

（1）百度地图的地理经纬度检索使用 GeoCoder 对象，需要事先声明，在使用前也要先实例化。

（2）调用 GeoCoder 对象的 geocode()方法由地址名检索其经纬度。

（3）reverseGeoCode()方法进行由经纬度到地址的反向搜索。

（4）地理经纬度检索的结果在 GeoCodeResult 类型的参数中返回，若返回结果为 SearchResult.ERRORNO.NO_ERROR 表示正确无误。

（5）将 icon_marka.png 图片资源以地图覆盖物的形式添加到地图上，用于标识检索出的地点。

9.3.3 运行效果

运行程序，输入"南京"市内一个地址"秦淮区中山东路 56 号"，点击"检索"，程序定位到地图上的该点，并显示对应的经纬度值，如图 9.29 所示。

反之，将这组经纬度数值输入文本框（以逗号分隔），点"检索"，程序同样会定位到该地点，但显示的是该地点的地址文字，如图 9.30 所示。

图 9.29　由地址名查经纬度　　　　　图 9.30　由经纬度值反查地址

9.4 Poi 检索

Poi 检索是地图应用最有用、最普及的功能。当人们来到一个陌生的城市，一般都急需一个旅行攻略，以便知道周边都有哪些美食店、商场、娱乐城、景点公园，等等。根据需求类型搜索地点，就是 Poi 的最大作用。Poi 还可以与其他检索类型（如公交路线检索）相结合，为用户提供更大的便利。

【例 9.3】运用 Poi 来检索南京市内的书店。

9.4.1 添加类库

百度地图新版 SDK 将地图开发中最常用的 Overlay 通用库的源码公开，而原本隶属于该库的所有类不再随 SDK 提供，开发者要使用这些类必须到百度官方提供的 Demo 项目工程中手工将此库开源后的包目录分离出来，再复制到项目 app\src\main\java\com 目录下，才能正常使用。如图 9.31 所示。

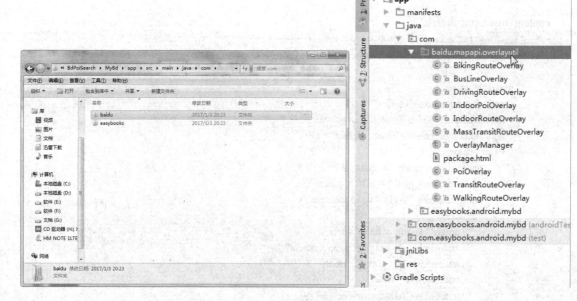

图 9.31　添加类库

为了方便读者，本书提供所有程序实例的源代码，读者可直接到项目工程中去获取 Overlay 通用库包，复制到自己的项目中使用。

9.4.2 设计界面

本项目的界面设计视图如图 9.32 所示。

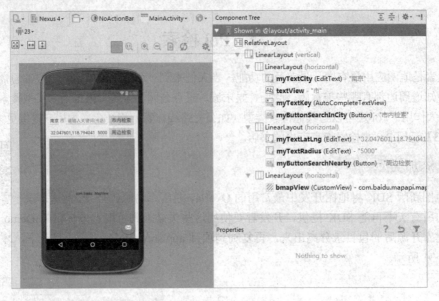

图9.32 Poi 检索设计视图

content_main.xml 源码如下：

```xml
<?xml version="1.0" encoding="utf-8"?>
<RelativeLayout xmlns:android="http://schemas.android.com/apk/res/android"
    ...>
    <LinearLayout
        android:orientation="vertical"
        android:layout_width="match_parent"
        android:layout_height="match_parent">
        <LinearLayout
            android:orientation="horizontal"
            android:layout_width="match_parent"
            android:layout_height="wrap_content">
            <EditText
                android:layout_width="45dp"
                android:layout_height="wrap_content"
                android:id="@+id/myTextCity"
                android:text="南京"
                android:textAlignment="center" />
            <TextView
                android:layout_width="wrap_content"
                android:layout_height="wrap_content"
                android:text="市"
                android:id="@+id/textView"
                android:textSize="@dimen/abc_text_size_title_material" />
            <AutoCompleteTextView
                android:layout_width="180dp"
                android:layout_height="wrap_content"
                android:id="@+id/myTextKey"
                android:textAlignment="center"
                android:hint="请输入关键词(书店)" />
            <Button
```

```xml
            android:layout_width="match_parent"
            android:layout_height="wrap_content"
            android:text="市内检索"
            android:id="@+id/myButtonSearchInCity"
            android:textSize="@dimen/abc_text_size_title_material" />
    </LinearLayout>
    <LinearLayout
        android:orientation="horizontal"
        android:layout_width="match_parent"
        android:layout_height="wrap_content">
        <EditText
            android:layout_width="195dp"
            android:layout_height="wrap_content"
            android:id="@+id/myTextLatLng"
            android:textAlignment="center"
            android:text="32.047601,118.794041" />
        <EditText
            android:layout_width="50dp"
            android:layout_height="wrap_content"
            android:id="@+id/myTextRadius"
            android:textAlignment="center"
            android:text="5000" />
        <Button
            android:layout_width="match_parent"
            android:layout_height="wrap_content"
            android:text="周边检索"
            android:id="@+id/myButtonSearchNearby"
            android:textSize="@dimen/abc_text_size_title_material" />
    </LinearLayout>
    <LinearLayout
        android:orientation="horizontal"
        android:layout_width="match_parent"
        android:layout_height="match_parent">
        <com.baidu.mapapi.map.MapView
            android:id="@+id/bmapView"
            android:layout_width="fill_parent"
            android:layout_height="fill_parent"
            android:clickable="true"/>
    </LinearLayout>
  </LinearLayout>
</RelativeLayout>
```

其中，输入检索关键词的文本框使用了 AutoCompleteTextView 控件，这是一个扩展功能的 TextView 控件，当用户在其中输入关键词时，它会自动弹出建议列表供用户选择匹配项。

将 icon_geo.png 图片（见图9.33）预先复制到项目 app\src\main\res\drawable 目录下。

图9.33 需要准备的图片资源

9.4.3 功能实现

本程序功能需要 PoiSearch 与 SuggestionSearch 相配合，首先由 PoiSearch 的 searchInCity()检索出市内所有符合要求的 Poi 地点，再由 SuggestionSearch 给出建议列表，列表信息的获取在 onGetSuggestionResult()方法中。Poi 信息的进一步处理在 onGetPoiResult()方法中实现，要根据检索类型（市内、周边）的不同分别进行处理。

MainActivity.java 代码如下：

```java
package com.easybooks.android.mybd;
import ...
public class MainActivity extends AppCompatActivity {
    private MapView myMap = null;
    private BaiduMap myBaiduMap = null;
    private EditText myCity;
    private AutoCompleteTextView myKey;
    private Button mySearchInCity;
    private EditText myLatLng;
    private EditText myRadius;
    private Button mySearchNearby;
    private PoiSearch myPoiSearch;                              //(1)
    private SuggestionSearch mySugSearch;                       //(2)
    private int searchType = 1;                                 //检索的类型（1:市内;2:周边）
    private LatLng center;                                      //中心点经纬度
    private int radius;                                         //检索半径
    private int loadIndex = 0;
    @Override
    protected void onCreate(Bundle savedInstanceState) {
        super.onCreate(savedInstanceState);
        //在使用 SDK 各组件之前初始化 context 信息，传入 ApplicationContext
        //注意该方法要在 setContentView 方法之前实现
        SDKInitializer.initialize(getApplicationContext());
        setContentView(R.layout.activity_main);
        //获取地图控件引用
        myMap = (MapView) findViewById(R.id.bmapView);
        myBaiduMap = myMap.getMap();
        //初始化
        findViews();
        myPoiSearch = PoiSearch.newInstance();
        mySugSearch = SuggestionSearch.newInstance();
        Toolbar toolbar = (Toolbar) findViewById(R.id.toolbar);
        ...
        //单击"市内检索"按钮的事件方法代码
        mySearchInCity.setOnClickListener(new View.OnClickListener() {
            @Override
            public void onClick(View v) {
                searchType = 1;
                String city = myCity.getText().toString();      //城市名
                String key = myKey.getText().toString();        //检索关键词
                myPoiSearch.searchInCity((new PoiCitySearchOption()).city(city).keyword(key).pageNum(loadIndex));  //(3)
            }
```

```java
        });
        //单击"周边检索"按钮的事件方法代码
        mySearchNearby.setOnClickListener(new View.OnClickListener() {
            @Override
            public void onClick(View v) {
                searchType = 2;
                String key = myKey.getText().toString();                    //检索关键词
                String[] laton = myLatLng.getText().toString().split(",");
                String lat = laton[0];                                       //经度
                String lon = laton[1];                                       //纬度
                center = new LatLng(Float.valueOf(lat), Float.valueOf(lon)); //中心点
                radius = Integer.valueOf(myRadius.getText().toString());     //搜索半径
                myPoiSearch.searchNearby(new PoiNearbySearchOption().keyword(key).sortType(PoiSortType.
distance_from_near_to_far).location(center).radius(radius).pageNum(loadIndex));    //(4)
            }
        });
        //回调方法获取并处理 Poi 检索的结果
        myPoiSearch.setOnGetPoiSearchResultListener(new OnGetPoiSearchResultListener() {
            @Override
            public void onGetPoiResult(PoiResult poiResult) {                //(5)
                if (poiResult.error == SearchResult.ERRORNO.NO_ERROR) {
                    myBaiduMap.clear();
                    PoiOverlay overlay = new MyPoiOverlay(myBaiduMap);       //(6)
                    myBaiduMap.setOnMarkerClickListener(overlay);
                    overlay.setData(poiResult);
                    overlay.addToMap();
                    overlay.zoomToSpan();
                    //如果是周边检索,需要对检索的范围进行绘制
                    if (searchType == 2) {
                        BitmapDescriptor centerBitmap = BitmapDescriptorFactory.fromResource (R.drawable.icon_geo);
                        MarkerOptions ooMarker = new MarkerOptions().position(center).icon(centerBitmap);
                        myBaiduMap.addOverlay(ooMarker);                     //在中心点添加图标
                        OverlayOptions ooCircle = new CircleOptions().fillColor(0xCCCCCC00). center(center).stroke(new Stroke(5, 0xFFFF00FF)).radius(radius);           //给检索区域(圆形)填充色彩
                        myBaiduMap.addOverlay(ooCircle);
                    }
                    return;
                }
            }

            @Override
            public void onGetPoiDetailResult(PoiDetailResult poiDetailResult) {
                if (poiDetailResult.error == SearchResult.ERRORNO.NO_ERROR) {
                    Toast.makeText(MainActivity.this, poiDetailResult.getName() + ": " + poiDetailResult.getAddress(), Toast.LENGTH_LONG).show();
                }
            }

            @Override
            public void onGetPoiIndoorResult(PoiIndoorResult poiIndoorResult) {
```

```java
            }
        });
        //回调方法获取并处理检索建议的结果
        mySugSearch.setOnGetSuggestionResultListener(new OnGetSuggestionResultListener() {
            @Override
            public void onGetSuggestionResult(SuggestionResult suggestionResult) {
                if (suggestionResult == null || suggestionResult.getAllSuggestions() == null) {
                    return;
                }
                List<String> suggest = new ArrayList<String>();
                for (SuggestionResult.SuggestionInfo info : suggestionResult.getAllSuggestions()) {
                    if (info.key != null) {
                        suggest.add(info.key);
                    }
                }
                ArrayAdapter<String> sugAdapter = new ArrayAdapter<String>(MainActivity.this, android.R.layout.simple_dropdown_item_1line, suggest);
                myKey.setAdapter(sugAdapter);
                sugAdapter.notifyDataSetChanged();
            }
        });
        //当输入关键词变化时，动态更新建议列表
        myKey.addTextChangedListener(new TextWatcher() {
            @Override
            public void beforeTextChanged(CharSequence s, int start, int count, int after) {
            }

            @Override
            public void onTextChanged(CharSequence s, int start, int before, int count) {
                //使用检索建议服务获取建议列表，结果在 onGetSuggestionResult()中更新
                if (s.length() > 0)
                    mySugSearch.requestSuggestion((new SuggestionSearchOption()).keyword(s.toString()).city(myCity.getText().toString()));
            }

            @Override
            public void afterTextChanged(Editable s) {            }
        });
    }

    private class MyPoiOverlay extends PoiOverlay {                              //(6)
        public MyPoiOverlay(BaiduMap baiduMap) {
            super(baiduMap);
        }

        @Override
        public boolean onPoiClick(int index) {
            super.onPoiClick(index);
            PoiInfo poi = getPoiResult().getAllPoi().get(index);
            myPoiSearch.searchPoiDetail((new PoiDetailSearchOption()).poiUid(poi.uid));
            return true;
        }
    }
```

```
    }
    private void findViews() {    …    }
    …
    @Override
    protected void onDestroy() {
        myPoiSearch.destroy();
        mySugSearch.destroy();
        super.onDestroy();
        …
    }
    …
}
```

其中：

（1）Poi 检索使用 PoiSearch 对象，需要事先声明和实例化。

（2）检索关键词建议使用 SuggestionSearch 对象，也要声明、实例化。

（3）调用 PoiSearch 对象的 searchInCity()方法实现在市内检索符合关键词要求的 Poi 地点。

（4）调用 PoiSearch 对象的 searchNearby()方法实现在指定半径区域范围内进行周边检索，sortType(PoiSortType.distance_from_near_to_far)指明搜索的顺序为由近及远。

（5）Poi 检索的结果集由 PoiResult 参数返回，判断检索是否成功的标准与前面介绍的几种检索完全一样。

（6）自定义 MyPoiOverlay 类继承自系统 Overlay 库（9.4.1 节添加）的 PoiOverlay，重写该类的 onPoiClick 方法，实现当用户点击地图上的标识时显示该地点的详细信息，详细信息由 PoiSearch 对象的 searchPoiDetail()方法获得，在回调函数 onGetPoiDetailResult 中处理。

9.4.4 运行效果

运行程序，输入检索关键词"新华书店"，点"市内检索"，程序在地图上标识出南京市内所有的新华书店分店所在位置，如图 9.34 所示。

指定一个经纬度，输入搜索半径 5000，点"周边检索"，程序在地图上标识出该点周围方圆 5km 范围内的新华书店分店，如图 9.35 所示。

图 9.34　在南京市内找新华书店　　　　图 9.35　指定地点周边 5km 范围内找新华书店

9.5 驾驶路径规划

T 在城市内驾车，司机经常需要路径导航功能，百度地图就提供了这样的能力，能够同时检索出到目的地的好几条路径，供司机选择最佳的路径。

【例 9.4】根据输入的起始地址和到达地址规划驾驶路径。

9.5.1 添加类库

该功能也要用到百度地图的 Overlay 库，需要用户手动添加，操作同 9.4.1 节，此处略。

9.5.2 设计界面

本项目的界面设计视图如图 9.36 所示。

图 9.36 驾驶路径规划设计视图

content_main.xml 源码如下：

```
<?xml version="1.0" encoding="utf-8"?>
<RelativeLayout xmlns:android="http://schemas.android.com/apk/res/android"
    ...>
    <LinearLayout
        android:orientation="vertical"
        android:layout_width="match_parent"
        android:layout_height="match_parent">
        <LinearLayout
            android:orientation="horizontal"
            android:layout_width="match_parent"
            android:layout_height="wrap_content"
            android:gravity="center_horizontal">
            <LinearLayout
                android:orientation="vertical"
                android:layout_width="wrap_content"
                android:layout_height="match_parent">
                <LinearLayout
                    android:orientation="horizontal"
```

```xml
            android:layout_width="wrap_content"
            android:layout_height="wrap_content">
            <TextView
                android:layout_width="wrap_content"
                android:layout_height="wrap_content"
                android:text="起    点"
                android:id="@+id/textView"
                android:textSize="@dimen/abc_text_size_title_material" />
            <EditText
                android:layout_width="190dp"
                android:layout_height="wrap_content"
                android:id="@+id/myTextStart"
                android:text="红山动物园"
                android:textAlignment="center" />
        </LinearLayout>
        <LinearLayout
            android:orientation="horizontal"
            android:layout_width="wrap_content"
            android:layout_height="wrap_content">
            <TextView
                android:layout_width="wrap_content"
                android:layout_height="wrap_content"
                android:text="终    点"
                android:id="@+id/textView2"
                android:textSize="@dimen/abc_text_size_title_material" />
            <EditText
                android:layout_width="190dp"
                android:layout_height="wrap_content"
                android:id="@+id/myTextEnd"
                android:text="新街口"
                android:textAlignment="center" />
        </LinearLayout>
    </LinearLayout>
    <LinearLayout
        android:orientation="vertical"
        android:layout_width="match_parent"
        android:layout_height="match_parent">
        <Button
            android:layout_width="match_parent"
            android:layout_height="match_parent"
            android:text="驾车规划"
            android:id="@+id/myButtonDrivePlan"
            android:textSize="@dimen/abc_text_size_title_material" />
    </LinearLayout>
</LinearLayout>
<RelativeLayout
    xmlns:android="http://schemas.android.com/apk/res/android"
    android:layout_width="match_parent"
    android:layout_height="match_parent">
    <com.baidu.mapapi.map.MapView
        android:id="@+id/bmapView"
        android:layout_width="fill_parent"
        android:layout_height="fill_parent"
```

```xml
                    android:clickable="true"/>
                <LinearLayout
                    android:orientation="horizontal"
                    android:layout_width="wrap_content"
                    android:layout_height="wrap_content"
                    android:layout_alignParentBottom="true"
                    android:gravity="center_horizontal"
                    android:layout_marginBottom="40dp"
                    android:layout_centerHorizontal="true">
                    <Button
                        android:layout_width="wrap_content"
                        android:layout_height="wrap_content"
                        android:id="@+id/myButtonPre"
                        android:background="@drawable/pre"
                        android:onClick="onNodeClick"
                        android:visibility="invisible" />
                    <Button
                        android:layout_width="wrap_content"
                        android:layout_height="wrap_content"
                        android:id="@+id/myButtonNext"
                        android:background="@drawable/next"
                        android:onClick="onNodeClick"
                        android:visibility="invisible" />
                </LinearLayout>
            </RelativeLayout>
    </LinearLayout>
</RelativeLayout>
```

将 next.png、popup.png、pre.png 图片（见图9.37）预先复制到项目 app\src\main\res\drawable 目录下。

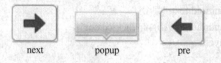

图 9.37 需要准备的图片资源

为了提供让用户选择不同路径的功能，在项目 app\src\main\res\layout 下创建两个文件 route_dialog.xml 和 route_item.xml，用于实现弹出供用户选择路径条目的对话框。

route_dialog.xml 源码如下：

```xml
<?xml version="1.0" encoding="utf-8"?>
<LinearLayout xmlns:android="http://schemas.android.com/apk/res/android"
            android:orientation="vertical"
            android:layout_width="match_parent"
            android:layout_height="match_parent">
    <TextView
            android:layout_width="wrap_content"
            android:layout_height="wrap_content"
            android:text="请点选所需路径，在地图上标示"
            android:textColor="#00ffff"
            android:padding="15dp"/>
    <ListView
            android:id="@+id/routeList"
            android:layout_width="fill_parent"
```

```xml
                android:layout_height="fill_parent"
                android:padding="15dp"
                android:divider="#00ffff"
                android:dividerHeight="1px"
                android:background="@color/background_floating_material_dark">
    </ListView>
</LinearLayout>
```

这里使用 ListView 控件实现对话框选择条目的列表。

Route_item.xml 源码如下：

```xml
<?xml version="1.0" encoding="utf-8"?>
<LinearLayout xmlns:android="http://schemas.android.com/apk/res/android"
                android:orientation="vertical"
                android:layout_width="match_parent"
                android:layout_height="match_parent">
    <TextView android:layout_width="match_parent"
                android:layout_height="match_parent"
                android:id="@+id/routeSeq"
                />
    <TextView android:layout_width="match_parent"
                android:layout_height="match_parent"
                android:id="@+id/lightNum"/>
    <TextView android:layout_width="match_parent"
                android:layout_height="match_parent"
                android:id="@+id/blockDis"/>
</LinearLayout>
```

该页面上定义了 3 个 TextView，分别用于显示每条路径的序号、红绿灯数及拥堵距离。

9.5.3 功能实现

创建 RouteAdapter 类，RouteAdapter.java 代码如下：

```java
package com.easybooks.android.mybd;
import ...
public class RouteAdapter extends BaseAdapter {
    private List<? extends RouteLine> routeLines;
    private LayoutInflater layoutInflater;
    private Type mtype;
    public RouteAdapter(Context context, List<? extends RouteLine> routeLines, Type type) {
        this.routeLines = routeLines;
        layoutInflater = LayoutInflater.from(context);
        mtype = type;
    }
    @Override
    public int getCount() {
        return routeLines.size();
    }
    @Override
    public Object getItem(int position) {
        return position;
    }
    @Override
```

```java
        public long getItemId(int position) {
            return position;
        }
        @Override
        public View getView(int position, View convertView, ViewGroup parent) {
            NodeViewHolder holder;
            if (convertView == null) {
                convertView = layoutInflater.inflate(R.layout.route_item, null);
                holder = new NodeViewHolder();
                holder.seq = (TextView) convertView.findViewById(R.id.routeSeq);
                holder.num = (TextView) convertView.findViewById(R.id.lightNum);
                holder.dis = (TextView) convertView.findViewById(R.id.blockDis);
                convertView.setTag(holder);
            } else {
                holder = (NodeViewHolder) convertView.getTag();
            }
            switch (mtype) {
                case WALKING_ROUTE:          //步行
                case TRANSIT_ROUTE:          //公交
                case MASS_TRANSIT_ROUTE:     //跨城
                case DRIVING_ROUTE:          //驾车
                    DrivingRouteLine drivingRouteLine = (DrivingRouteLine) routeLines.get(position);
                    holder.seq.setText("路径" + (position + 1));
                    holder.num.setText("红绿灯数：" + drivingRouteLine.getLightNum());
                    holder.dis.setText("拥堵距离为：" + drivingRouteLine.getCongestionDistance() + "米");
                    break;
                case INDOOR_ROUTE:           //室内
                case BIKING_ROUTE:           //骑行
                default:
                    break;
            }
            return convertView;
        }
        private class NodeViewHolder {
            private TextView seq;
            private TextView num;
            private TextView dis;
        }
        public enum Type {
            WALKING_ROUTE,               //步行
            TRANSIT_ROUTE,               //公交
            MASS_TRANSIT_ROUTE,          //跨城
            DRIVING_ROUTE,               //驾车
            INDOOR_ROUTE,                //室内
            BIKING_ROUTE                 //骑行
        }
    }
}
```

这是一个通用的适配器类，可根据出行方式的不同显示不同类型路径的条目信息，支持步行、公交、跨城、驾车、室内和骑行等多种方式，本项目只是实现驾车路径规划，故只要在 switch 语句的"case DRIVING_ROUTE:"分支后面编写代码，至于其他出行方式，可以先留着空位，今后逐步扩充即可。

主程序功能在 MainActivity 中实现，当单击"驾车规划"按钮时，由界面文本框获取起终点信息，通过 drivingSearch()方法传递给 RoutePlanSearch 对象去检索（加黑语句），检索返回的结果在 onGetDrivingRouteResult()方法中做进一步的处理。

MainActivity.java 代码如下：

```java
package com.easybooks.android.mybd;
import ...
public class MainActivity extends AppCompatActivity {
    private MapView myMap = null;
    private BaiduMap myBaiduMap = null;
    private EditText myStart;
    private EditText myEnd;
    private Button myDrivePlan;
    private Button myPre;
    private Button myNext;
    private RoutePlanSearch myRoutePlanSearch;          //(1)
    private RouteLine route;                            //(2)
    private int nodeIndex = -1;                         //节点索引，供浏览节点时使用
    @Override
    protected void onCreate(Bundle savedInstanceState) {
        super.onCreate(savedInstanceState);
        //在使用 SDK 各组件之前初始化 context 信息，传入 ApplicationContext
        //注意该方法要在 setContentView 方法之前实现
        SDKInitializer.initialize(getApplicationContext());
        setContentView(R.layout.activity_main);
        //获取地图控件引用
        myMap = (MapView) findViewById(R.id.bmapView);
        myBaiduMap = myMap.getMap();
        //初始化
        findViews();
        myRoutePlanSearch = RoutePlanSearch.newInstance();
        Toolbar toolbar = (Toolbar) findViewById(R.id.toolbar);
        ...
        //单击"驾车规划"按钮的事件方法代码
        myDrivePlan.setOnClickListener(new View.OnClickListener() {
            @Override
            public void onClick(View v) {
                //重置浏览节点的路线数据
                route = null;
                myPre.setVisibility(View.INVISIBLE);
                myNext.setVisibility(View.INVISIBLE);
                myBaiduMap.clear();
                //设置起终点信息
                PlanNode stNode = PlanNode.withCityNameAndPlaceName("南京", myStart.getText().toString());
                PlanNode enNode = PlanNode.withCityNameAndPlaceName("南京", myEnd.getText().toString());
                //开始检索路径规划
                myRoutePlanSearch.drivingSearch((new DrivingRoutePlanOption()).from(stNode).to(enNode));  //(3)
            }
```

```java
        });
        //回调方法获取并处理路径规划检索的结果
        myRoutePlanSearch.setOnGetRoutePlanResultListener(new OnGetRoutePlanResultListener() {
                                                                            //(4)
            @Override
            public void onGetWalkingRouteResult(WalkingRouteResult walkingRouteResult) {
                //获取步行路径、处理...
            }
            @Override
            public void onGetTransitRouteResult(TransitRouteResult transitRouteResult) {
                //获取公交路径、处理...
            }
            @Override
            public void onGetMassTransitRouteResult(MassTransitRouteResult massTransitRouteResult) {
                //获取跨城路径、处理...
            }
            @Override
            public void onGetDrivingRouteResult(final DrivingRouteResult drivingRouteResult) {
                //获取驾车路径、处理...
                if (drivingRouteResult.error == SearchResult.ERRORNO.NO_ERROR) {
                    nodeIndex = -1;
                    if (drivingRouteResult.getRouteLines().size() > 1) {    //(5)
                        MyRouteDlg myRouteDlg = new MyRouteDlg(MainActivity.this, drivingRouteResult.getRouteLines(), RouteAdapter.Type.DRIVING_ROUTE);
                        myRouteDlg.setOnItemInDlgClickListener(new OnItemInDlgClickListener() {
                            @Override
                            public void onItemClick(int position) {
                                route = drivingRouteResult.getRouteLines().get(position);
                                DrivingRouteOverlay overlay = new DrivingRouteOverlay(myBaiduMap);
                                myBaiduMap.setOnMarkerClickListener(overlay);
                                overlay.setData(drivingRouteResult.getRouteLines().get(position));
                                overlay.addToMap();
                                overlay.zoomToSpan();
                            }
                        });
                        myRouteDlg.show();
                    } else if (drivingRouteResult.getRouteLines().size() == 1) {
                        route = drivingRouteResult.getRouteLines().get(0);
                        DrivingRouteOverlay overlay = new DrivingRouteOverlay(myBaiduMap);
                        myBaiduMap.setOnMarkerClickListener(overlay);
                        overlay.setData(drivingRouteResult.getRouteLines().get(0));
                        overlay.addToMap();
                        overlay.zoomToSpan();
                        myPre.setVisibility(View.VISIBLE);
                        myNext.setVisibility(View.VISIBLE);
                    }
                }
            }
            @Override
            public void onGetIndoorRouteResult(IndoorRouteResult indoorRouteResult) {
                //获取室内路径、处理...
```

```java
            }
            @Override
            public void onGetBikingRouteResult(BikingRouteResult bikingRouteResult) {
                //获取骑行路径、处理...
            }
        });
    }
    //供路线选择的对话框
    public class MyRouteDlg extends Dialog {
        private List<? extends RouteLine> mRouteLines;
        private ListView mRouteList;
        private RouteAdapter mRouteAdapter;
        OnItemInDlgClickListener onItemInDlgClickListener;
        public MyRouteDlg(Context context, int theme) {
            super(context, theme);
        }
        public MyRouteDlg(Context context, List<? extends RouteLine> routeLines, RouteAdapter.Type type) {
            this(context, 0);
            mRouteLines = routeLines;
            mRouteAdapter = new RouteAdapter(context, mRouteLines, type);
            requestWindowFeature(Window.FEATURE_NO_TITLE);
        }
        @Override
        protected void onCreate(Bundle savedInstanceState) {
            super.onCreate(savedInstanceState);
            setContentView(R.layout.route_dialog);
            mRouteList = (ListView) findViewById(R.id.routeList);
            mRouteList.setAdapter(mRouteAdapter);
            mRouteList.setOnItemClickListener(new AdapterView.OnItemClickListener() {
                @Override
                public void onItemClick(AdapterView<?> parent, View view, int position, long id) {
                    onItemInDlgClickListener.onItemClick(position);
                    myPre.setVisibility(View.VISIBLE);
                    myNext.setVisibility(View.VISIBLE);
                    dismiss();
                }
            });
        }

        public void setOnItemInDlgClickListener(OnItemInDlgClickListener itemListener) {
            onItemInDlgClickListener = itemListener;
        }
    }
    //响应对话框中的条目点击
    public interface OnItemInDlgClickListener {
        void onItemClick(int position);
    }
    //单击前后箭头按钮浏览节点详细信息
    public void onNodeClick(View v) {
        LatLng nodeLocation = null;
        String nodeTitle = null;
```

```
            Object step = null;
                //非跨城综合交通
                if (route == null || route.getAllStep() == null) {
                    return;
                }
                if (nodeIndex == -1 && v.getId() == R.id.myButtonPre) {
                    return;
                }
                //设置节点索引
                if (v.getId() == R.id.myButtonNext) {
                    if (nodeIndex < route.getAllStep().size() -1) {
                        nodeIndex++;
                    } else {
                        return;
                    }
                } else if (v.getId() == R.id.myButtonPre) {
                    if (nodeIndex > 0) {
                        nodeIndex--;
                    } else {
                        return;
                    }
                }
                //获取节点信息
                step = route.getAllStep().get(nodeIndex);
                nodeLocation = ((DrivingRouteLine.DrivingStep) step).getEntrance().getLocation();
                nodeTitle = ((DrivingRouteLine.DrivingStep) step).getInstructions();
                if (nodeLocation == null || nodeTitle == null) {
                    return;
                }
                //移动节点至中心
                myBaiduMap.setMapStatus(MapStatusUpdateFactory.newLatLng(nodeLocation));
                //弹出显示节点信息的标签
                TextView popupText = new TextView(MainActivity.this);
                popupText.setBackgroundResource(R.drawable.popup);
                popupText.setTextColor(0xFF000000);
                popupText.setText(nodeTitle);
                myBaiduMap.showInfoWindow(new InfoWindow(popupText, nodeLocation, 0));
        }
        private void findViews() {   …   }
            …
        @Override
        protected void onDestroy() {
            myRoutePlanSearch.destroy();
            super.onDestroy();
            …
        }
            …
    }
```

其中：

（1）驾车路径规划检索使用 RoutePlanSearch 对象，要声明和实例化。

(2) RouteLine 对象用于存储某条路径的信息。

(3) 调用 RoutePlanSearch 对象的 drivingSearch()方法检索符合条件的路径,需要给出起终点作为参数。

(4) setOnGetRoutePlanResultListener 回调方法可支持多种不同出行方式的路径规划,本项目要做的是驾车路径规划,故只要重写 onGetDrivingRouteResult()方法即可,但其他出行方式所对应的处理方法也要写出其方法体(虽然实际上用不着,但作为接口实现不可或缺!)。

(5) 检索到符合条件的驾车路径通过 DrivingRouteResult 类的 getRouteLines()方法获取,若获取的路径数大于 1,则弹出对话框让用户选择。若只有唯一的路径,则直接显示结果。

9.5.4 运行效果

运行程序,在南京市内设置从红山动物园(起点)至新街口(终点)的路径,点"驾车规划",在弹出路径选择对话框中,我们选择沿途红绿灯数最少的"路径 3",如图 9.38 所示。

接着,程序在地图上标示出该路径,自动标识出起(终)点,如图 9.39 所示。

点击地图下方的前后箭头,可从弹出的消息文本中查看该路径上沿途各关键地点的导航信息,如图 9.40 所示。

图 9.38 选择路径　　　　图 9.39 在地图上标示路径　　　　图 9.40 开始导航

9.6 公交线路查询

本项目主要实现查询市内某路公交车沿途各站点信息的功能,同样需要使用到 Overlay 库。

【例 9.5】根据输入的城市和公交车线路编号,查询公交车沿途站点及其连线图。

9.6.1 添加类库

操作同 9.4.1 节,此处略。

9.6.2 设计界面

本项目界面设计视图如图 9.41 所示。

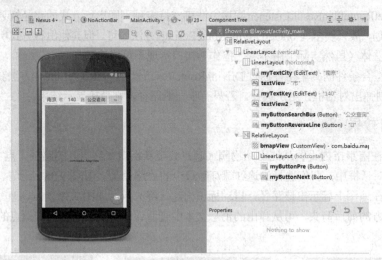

图 9.41 公交线路查询设计视图

content_main.xml 代码如下：

```xml
<?xml version="1.0" encoding="utf-8"?>
<RelativeLayout xmlns:android="http://schemas.android.com/apk/res/android"
      ...>
    <LinearLayout
        android:orientation="vertical"
        android:layout_width="match_parent"
        android:layout_height="match_parent">
        <LinearLayout
            android:orientation="horizontal"
            android:layout_width="match_parent"
            android:layout_height="wrap_content"
            android:gravity="center_horizontal">
            <EditText
                android:layout_width="60dp"
                android:layout_height="wrap_content"
                android:id="@+id/myTextCity"
                android:text="南京"
                android:textAlignment="center"
                android:textSize="@dimen/abc_text_size_large_material" />
            <TextView
                android:layout_width="wrap_content"
                android:layout_height="wrap_content"
                android:text="市"
                android:id="@+id/textView"
                android:textSize="@dimen/abc_text_size_title_material" />
            <EditText
                android:layout_width="80dp"
                android:layout_height="wrap_content"
```

```xml
            android:id="@+id/myTextKey"
            android:text="140"
            android:textAlignment="center"
            android:textSize="@dimen/abc_text_size_large_material" />
    <TextView
            android:layout_width="wrap_content"
            android:layout_height="wrap_content"
            android:text="路"
            android:id="@+id/textView2"
            android:textSize="@dimen/abc_text_size_title_material" />
    <Button
            android:layout_width="wrap_content"
            android:layout_height="wrap_content"
            android:text="公交查询"
            android:id="@+id/myButtonSearchBus"
            android:textSize="@dimen/abc_text_size_title_material" />
    <Button
            android:layout_width="wrap_content"
            android:layout_height="wrap_content"
            android:text="⇌"
            android:id="@+id/myButtonReverseLine"
            android:textSize="@dimen/abc_text_size_large_material" />
</LinearLayout>
<RelativeLayout
    xmlns:android="http://schemas.android.com/apk/res/android"
    android:layout_width="match_parent"
    android:layout_height="match_parent">
    <com.baidu.mapapi.map.MapView
        android:id="@+id/bmapView"
        android:layout_width="fill_parent"
        android:layout_height="fill_parent"
        android:clickable="true"/>
    <LinearLayout
        android:orientation="horizontal"
        android:layout_width="wrap_content"
        android:layout_height="wrap_content"
        android:layout_alignParentBottom="true"
        android:gravity="center_horizontal"
        android:layout_marginBottom="40dp"
        android:layout_centerHorizontal="true">
        <Button
            android:layout_width="wrap_content"
            android:layout_height="wrap_content"
            android:id="@+id/myButtonPre"
            android:background="@drawable/pre"
            android:onClick="onNodeClick"
            android:visibility="invisible" />
        <Button
            android:layout_width="wrap_content"
            android:layout_height="wrap_content"
            android:id="@+id/myButtonNext"
```

```xml
                        android:background="@drawable/next"
                        android:onClick="onNodeClick"
                        android:visibility="invisible" />
                </LinearLayout>
            </RelativeLayout>
        </LinearLayout>
</RelativeLayout>
```

将 next.png、popup.png、pre.png 三张图片（见图 9.37）预先复制到项目 app\src\main\res\drawable 目录下。

9.6.3 功能实现

当用户单击"公交查询"按钮时，系统调用 PoiSearch 的 searchInCity()方法查找市内符合要求的 Poi（加黑语句），Poi 检索得到的结果先在 onGetPoiResult()方法中进行处理，找出公交线路类型的 Poi。接着调用"myReverseLine.performClick();"触发公交线路查询，调用 BusLineSearch 对象的 searchBusLine()方法进一步进行公交线路的详情搜索，搜索返回的结果在 onGetBusLineResult()方法中处理。

MainActivity.java 代码如下：

```java
package com.easybooks.android.mybd;
import ...
public class MainActivity extends AppCompatActivity {
    private MapView myMap = null;
    private BaiduMap myBaiduMap = null;
    private EditText myCity;
    private EditText myKey;
    private Button mySearchBus;
    private Button myReverseLine;
    private Button myPre;
    private Button myNext;
    private PoiSearch myPoiSearch;                        //(1)
    private BusLineSearch myBusLineSearch;                //(2)
    private List<String> busLineIDList = null;
    private BusLineResult route = null;           //保存公交线路数据的变量，供浏览节点时使用
    private int nodeIndex = -2;                   //节点索引，供浏览节点时使用
    private int busLineIndex = 0;
    @Override
    protected void onCreate(Bundle savedInstanceState) {
        super.onCreate(savedInstanceState);
        //在使用 SDK 各组件之前初始化 context 信息，传入 ApplicationContext
        //注意该方法要在 setContentView 方法之前实现
        SDKInitializer.initialize(getApplicationContext());
        setContentView(R.layout.activity_main);
        //获取地图控件引用
        myMap = (MapView) findViewById(R.id.bmapView);
        myBaiduMap = myMap.getMap();
        //初始化
        findViews();
        myPoiSearch = PoiSearch.newInstance();
        myBusLineSearch = BusLineSearch.newInstance();
```

```java
        busLineIDList = new ArrayList<String>();
        Toolbar toolbar = (Toolbar) findViewById(R.id.toolbar);
        ...
        //单击"公交查询"按钮的事件方法代码
        mySearchBus.setOnClickListener(new View.OnClickListener() {
            @Override
            public void onClick(View v) {
                busLineIDList.clear();
                busLineIndex = 0;
                myPre.setVisibility(View.INVISIBLE);
                myNext.setVisibility(View.INVISIBLE);
                //发起 poi 检索,从得到所有 poi 中找到公交线路类型的 poi,再使用该 poi 的 uid 进行公交详情搜索
                myPoiSearch.searchInCity((new PoiCitySearchOption()).city(myCity.getText().toString()).keyword(myKey.getText().toString()));
            }
        });
        //单击"⇆"按钮的事件方法代码
        myReverseLine.setOnClickListener(new View.OnClickListener() {
            @Override
            public void onClick(View v) {
                if (busLineIndex >= busLineIDList.size()) {
                    busLineIndex = 0;
                }
                if (busLineIndex >= 0 && busLineIndex < busLineIDList.size() && busLineIDList.size() > 0) {
                    myBusLineSearch.searchBusLine((new BusLineSearchOption().city(myCity.getText().toString()).uid(busLineIDList.get(busLineIndex))));    //(3)
                    busLineIndex++;
                }
            }
        });
        //回调方法获取并处理 Poi 检索的结果
        myPoiSearch.setOnGetPoiSearchResultListener(new OnGetPoiSearchResultListener() {    //(4)
            @Override
            public void onGetPoiResult(PoiResult poiResult) {
                if (poiResult.error == SearchResult.ERRORNO.NO_ERROR) {
                    //遍历所有 poi,找到类型为公交线路的 poi
                    busLineIDList.clear();
                    for (PoiInfo poi : poiResult.getAllPoi()) {
                        if (poi.type == PoiInfo.POITYPE.BUS_LINE || poi.type == PoiInfo.POITYPE.SUBWAY_LINE) {
                            busLineIDList.add(poi.uid);             //添加进公交线路列表
                        }
                    }
                    myReverseLine.performClick();
                    route = null;
                }
            }
            @Override
            public void onGetPoiDetailResult(PoiDetailResult poiDetailResult) {          }
            @Override
```

```java
            public void onGetPoiIndoorResult(PoiIndoorResult poiIndoorResult) {            }
        });
        //回调方法获取并处理公交线路检索的结果
        myBusLineSearch.setOnGetBusLineSearchResultListener(new OnGetBusLineSearchResultListener() {
            @Override
            public void onGetBusLineResult(BusLineResult busLineResult) {
                if (busLineResult.error == SearchResult.ERRORNO.NO_ERROR) {
                    myBaiduMap.clear();
                    route = busLineResult;
                    nodeIndex = -1;
                    BusLineOverlay overlay = new BusLineOverlay(myBaiduMap);
                    myBaiduMap.setOnMarkerClickListener(overlay);
                    overlay.removeFromMap();
                    overlay.setData(busLineResult);
                    overlay.addToMap();
                    overlay.zoomToSpan();
                    myPre.setVisibility(View.VISIBLE);
                    myNext.setVisibility(View.VISIBLE);
                    Toast.makeText(MainActivity.this, busLineResult.getBusLineName(), Toast.LENGTH_LONG).show();
                }
            }
        });
    }
    //单击前后箭头按钮浏览节点(公交站点)详细信息
    public void onNodeClick(View v) {
        if (nodeIndex < -1 || route == null || nodeIndex >= route.getStations().size()) {
            return;
        }
        TextView popupText = new TextView(this);
        popupText.setBackgroundResource(R.drawable.popup);
        popupText.setTextColor(0xff000000);
        //上一个节点
        if (myPre.equals(v) && nodeIndex > 0) {
            nodeIndex--;            //索引减
        }
        //下一个节点
        if (myNext.equals(v) && nodeIndex < (route.getStations().size() -1)) {
            nodeIndex++;            //索引加
        }
        if (nodeIndex >= 0) {
            //移动到指定索引的坐标
            myBaiduMap.setMapStatus(MapStatusUpdateFactory.newLatLng(route.getStations().get(nodeIndex).getLocation()));
            //弹出泡泡
            popupText.setText(route.getStations().get(nodeIndex).getTitle());
            myBaiduMap.showInfoWindow(new InfoWindow(popupText, route.getStations().get(nodeIndex).getLocation(), 10));
        }
    }
    private void findViews() {  …  }
```

```
        …
    @Override
    protected void onDestroy() {
        myPoiSearch.destroy();
        myBusLineSearch.destroy();
        super.onDestroy();
        …
    }
        …
}
```

其中：

（1）本项目功能要用到 Poi 检索，故要与 PoiSearch 对象配合使用。

（2）BusLineSearch 对象用于公交线路查询，需要声明和实例化。

（3）调用 BusLineSearch 对象的 searchBusLine()方法查询给定索引的公交线路。

（4）setOnGetBusLineSearchResultListener 回调方法可支持获取不同形式的 Poi 信息，本项目只需获得公交线路类型 Poi 的基本信息，故只要重写 onGetPoiResult()方法即可，但作为接口使用规范，其他两个处理方法（onGetPoiDetailResult 和 onGetPoiIndoorResult）也要写出其方法体。

9.6.4 运行效果

运行程序，填写南京市内 140 路公交，点"公交查询"，如图 9.42 所示，程序会在地图上标示出该条公交线路。

点地图下部的前后箭头，可翻看 140 路沿途各站的信息，如图 9.43 所示。

点"⇋"按钮，可查询出 140 路反向线路的信息，如图 9.44 所示。

图 9.42　查询 140 路公交线　　图 9.43　翻看 140 路沿途各站信息　　图 9.44　查看 140 路反向线路信息

9.7　高德地图开发

高德地图的开发与百度地图有很多相似之处，所不同的主要是接口 API，读者可比照本章百度地图开发的源代码，同时参考高德官网上的开发指南（见图 9.45）来实现与上述百度地图一模一样的应用功能。高德地图开发需要在 9.1.2 创建的 Android 项目 MyGd 的基础上做。

图 9.45　高德官网上的开发指南

高德地图与百度地图相比，在功能上大致有如下区别。
● 地图类型
增加了夜景地图 MAP_TYPE_NIGHT 类型，但没有城市热力图。
设置夜景模式的语句为：

aMap.setMapType(AMap.MAP_TYPE_NIGHT);　　　　　　//夜景地图，aMap 是地图控制器对象
● 区域检索
DistrictSearch 需要与 DistrictSearchQuery 类一起配合使用。
典型代码：

DistrictSearch search = new DistrictSearch(mContext);
DistrictSearchQuery query = new DistrictSearchQuery();
query.setKeywords("玄武区");　　　　　　　　　　　//传入关键字
search.setQuery(query);
search.searchDistrictAnsy();　　　　　　　　　　　//开始搜索

● 地理经纬度检索
使用 GeocodeSearch 类实现，需要与 GeocodeQuery/RegeocodeQuery 类配合。
调用 getFromLocationNameAsyn(GeocodeQuery geocodeQuery) 方法进行 Geo 搜索（地址名→经纬度）。
典型代码：

geocoderSearch = new GeocodeSearch(this);
GeocodeQuery query = new GeocodeQuery(name, "南京");
geocoderSearch.getFromLocationNameAsyn(query);

调用 getFromLocationAsyn()方法进行反 Geo 搜索（经纬度→地址名）。
典型代码：

RegeocodeQuery query = new RegeocodeQuery(latLonPoint, 200,GeocodeSearch.AMAP);
geocoderSearch.getFromLocationAsyn(query);

第一个参数表示一个经纬度对象，第二个参数表示范围多少米，第三个参数表示是火星坐标系还是 GPS 原生坐标系。

● Poi 检索

需要构造 PoiSearch.Query 对象，通过 PoiSearch.Query(String query, String ctgr, String city) 设置搜索条件，然后与 PoiSearch 对象配合使用。

典型代码：

```
query = new PoiSearch.Query("新华书店", "", "南京");
query.setPageSize(10);                    //设置每页最多返回多少条 poiitem
query.setPageNum(currentPage);            //设置查询页码
poiSearch = new PoiSearch(this, query);
poiSearch.searchPOIAsyn();
```

采用 Inputtips 与 InputtipsQuery 对象相配合的方式实现输入提示功能，作用等同于百度地图的 SuggestionSearch 类。

● 驾驶路径规划

使用 RouteSearch 类实现，通过 DriveRouteQuery 设置搜索条件，可设置如下参数。

fromAndTo：路径的起点终点。

mode：计算路径的模式，可选，默认为速度优先。

passedByPoints：途经点，可选。

avoidpolygons：避让区域，可选，支持 32 个避让区域，每个区域最多可有 16 个顶点。如果是四边形则有 4 个坐标点，如果是五边形则有 5 个坐标点。

avoidRoad：避让道路，只支持一条避让道路，避让区域和避让道路同时设置，只有避让道路生效。

fromAndTo 包含路径规划的起点和终点，drivingMode 表示驾车模式。

第三个参数表示途经点（最多支持 16 个），第四个参数表示避让区域（最多支持 32 个），第五个参数表示避让道路。

```
DriveRouteQuery query = new DriveRouteQuery(fromAndTo, drivingMode, null, null, "");
```

使用 RouteSearch 的 calculateRideRouteAsyn(RideRouteQuery query)方法进行驾车规划路径计算。

代码如下：

```
routeSearch.calculateDriveRouteAsyn(query);
```

● 公交线路查询

使用 BusLineSearch 类，调用其 searchBusLineAsyn()方法搜索公交线路，与百度相比不同在于：无须与 PoiSearch 结合使用。

典型代码：

```
busLineQuery = new BusLineQuery("140",SearchType.BY_LINE_NAME, "南京");
busLineQuery.setPageSize(10);
busLineQuery.setPageNumber(currentpage);
BusLineSearch busLineSearch = new BusLineSearch(this,busLineQuery);
busLineSearch.searchBusLineAsyn();
```

更多详细内容请读者自己上高德官网学习，限于篇幅，本章不再赘述。

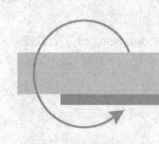

习题和实验

第 1 章　Android 概述

一、填空题

1. Android 是 Google 公司基于_____平台开发的手机及平板电脑的应用程序。
2. Android 的分层架构中，应用框架层使用_____语言开发。
3. 程序员编写 Android 应用程序时，主要调用_____层提供的接口实现。
4. 为了让程序员更加方便地运行调试程序，Android 提供了_____。
5. Android Studio 是一项全新的基于_____的 Android 开发环境，类似于_____插件。

二、选择题

1. Android Studio 软硬件要求（多选题）（　　）。
A．Windows 7（32 位）　　B．4GB 以上内存　　C．i7CPU　　D．固态硬盘
2. Android Studio 开发环境需要（　　）。
A．JDK
B．配置 JDK 的环境变量
C．Android Studio 安装成功
D．第一次运行配置完成

三、简答题

1. 当前主流的 Android 开发环境有哪两种？比较它们各自的利弊。
2. Android Studio 平台的特征和优势是什么？
3. Android Studio 开发环境安装的步骤有哪些？
4. Android Studio 安装的默认路径是什么？
5. Android Studio 平台应用程序创建和运行的步骤有哪些？

四、实验题

1. 安装 JDK（参考教材 1.3 节）

（1）建议从 Oracle 官网下载到最新版本的 JDK，网址为 http://www.oracle.com/technetwork/java/javase/downloads/index.html。

（2）安装下载 JDK 安装包。

（3）配置 Windows 环境变量，使后面安装的 Android Studio 能够找到 JDK。

2. 安装 Android Studio（参考教材 1.3 节）

（1）下载 Android Studio。

（2）安装 Android Studio，完成第一次启动工作。

第 2 章　Android 开发入门

一、填空题

1. Android 工程创建后，默认生成一个布局文件 activity_main.xml，该文件位于项目的＿＿＿＿＿＿＿文件夹中。默认生成一个 Activity 文件 MainActivity.java，该文件中会自动导入＿＿＿＿＿＿＿类。
2. 在 Android 应用程序中，界面布局文件在工程的＿＿＿＿＿＿＿文件夹，采用＿＿＿＿＿＿＿格式。
3. Android 应用程序的配置文件名称为＿＿＿＿＿＿＿。
4. 如果需要创建一个字符串资源，需要将字符串放在 res\values＿＿＿＿＿＿＿文件中。
5. 若要将 Android 程序发布到 Internet，需要将程序＿＿＿＿＿＿＿。

二、简答题

1. 创建第一个 Android 项目，说明它由哪几部分构成。
2. 说明 Android Studio 开发环境的特点。
3. 在仿真器中模拟运行这个 Android 项目，有条件的读者还可试着用自己的手机运行。
4. 修改 Android 项目，在界面上显示图片，并试着改变"Hello World!"文字的属性查看显示效果。
5. Android 程序修改属性有哪两种方式？各有什么优缺点？
6. Android 程序有哪两种事件处理机制？说说这两种机制各自的开发流程及利弊。
7. 总结本书的实例代码规范，为后面的学习做准备。

三、说明题

1. 请说明在创建 Android 工程时下面名称的意义。
（1）Application name
（2）Company Domain
（3）Package name
（4）Project location
（5）Activity Name
（6）Layout Name
（7）Title
（8）Menu Resource Name
2. 在创建 Android 工程时为什么需要选择应用程序要运行的平台？
3. Editor（编辑器）窗口功能有哪些？
4. 折叠线、标记栏、导航栏和状态栏的作用有哪些？
5. Android 工程结构是怎样的？
6. 模拟运行和真机运行有什么差别？
7. 在编辑 XML 界面文件时 Design 和 Text 方式有什么不同？
8. 配置字符串资源文件的作用是什么？

四、实验题

1. 创建第一个 Android 工程（参考教材 2.1 节）。
（1）创建 HelloWorld 工程。

(2) 了解 Android Studio 工程开发环境。
(3) 初步熟悉 Android Studio 工程结构。
(4) 模拟运行。
(5) 真机运行。
2. 修改 Hello World 程序（参考教材 2.2 节）。
(1) 修改页面、编写事件处理代码。
(2) 模拟运行。

第 3 章　Android 用户界面

一、选择题

1. Android 中有许多控件，这些控件都继承自（　　）类。
   A. Control　　　　　B. Window　　　　　C. TextView　　　　　D. View
2. "指定某控件左边"属性是（　　）。
   A. android:layout_alignleft　　　　　　B. android:layout_alignParentLeft
   C. android:layout_left　　　　　　　　 D. android:layout_toLeftOf
3. 显示文本最合适的控件是（　　）。
   A. ImageView　　　B. TextView　　　　C. EditText　　　　　D. Button
4. 用来显示图片的是（　　）控件。
   A. ImageView　　　B. TextView　　　　C. EditText　　　　　D. Button
5. （　　）属性用来表示引用图片的资源。
   A. text　　　　　　B. img　　　　　　　C. id　　　　　　　　D. src
6. （　　）方法可以用来获得进度条的当前进度值。
   A. getProgress()　　　　　　　　　　　B. setIndeterminate()
   C. setProgress()　　　　　　　　　　　D. incrementProgressBy()
7. ListView 是（　　）控件。
   A. 单选按钮　　　　B. 复选框　　　　　C. 列表　　　　　　　D. 下拉列表
8. 如果需要捕捉某个控件的事件，需要为该控件创建（　　）。
   A. 属性　　　　　　B. 方法　　　　　　C. 监听器　　　　　　D. 事件

二、简答题

1. 在 Android 的布局文件 activity_main.xml 中 "@+id/username" 与 "@id/username" 两者有何区别？
2. 控件的只读属性和是否可用属性有什么不同？
3. 字符显示和编辑控件功能有哪些差别？
4. 按钮和图像按钮控件功能有哪些差别？
5. 下拉列表和列表框功能有哪些差别？用户选择下拉列表和列表框的值如何得到？
6. 单选按钮为什么需要放在它的容器中？

三、实验题

1. 熟悉基本的界面控件（参考教材 3.2 节）。

（1）按照本节教材实例操作，模拟运行测试。
（2）修改实例有关的控件属性，观察界面效果。
（3）修改实例有关的控件事件属性和事件代码，模拟运行观察效果。
2．熟悉高级控件应用（参考教材 3.4 节）。
（1）按照本节教材实例操作，模拟运行测试。
（2）修改实例有关的控件属性，观察界面效果。
3．应用设计。
（1）应用本章内容设计学生基本信息输入页面，单击"提交"命令按钮，根据按钮单击属性指定的方法编写代码，将用户输入的信息用对话框显示出来。
（2）应用本章内容设计图书信息输入页面，单击"提交"命令按钮，编写按钮点击事件监听器代码，将用户输入的信息用对话框显示出来。
（3）分别用网页浏览控件、滚动预览控件、照片查看器控件、条类控制器控件等控件及它们的组合设计小应用实例。

第 4 章　用户界面布局

一、填空题

1．创建 Android 程序时，默认使用的布局是_____。
2．在相对布局中，_____属性确定横向相对布局；_____属性表示"是否跟父布局左对齐"；_____属性表示"与指定控件右对齐"。
3．表格布局可以包含多行，_____代表是一行。
4．帧布局中可以添加多个控件，这些控件会重叠在屏幕_____显示。
5．根据控件的_____属性获取控件的对象。如果控件的 id 设置为 myButton，那么调用方法 findViewById()时，引用该控件的参数应为_____。

二、选择题

1．(　　)属性可以指定"在指定控件左边"。
　A．android:layout_alignleft　　　　　　B．android:layout_alignParentLeft
　C．android:layout_left　　　　　　　　D．android:layout_toLeftOf
2．在相对布局中，(　　)属性确定"是否跟父布局底部对齐"。
　A．android:layout_alignBottom　　　　B．android:layout_alignParentBottom
　C．android:layout_alignBaseline　　　　D．android:layout below
3．在相对布局中，(　　)属性指定一个控件位于引用控件的左侧。
　A．android:layout toParentLeftOf　　　B．android:layout_alignParentLeft
　C．android:layout_alignLeft　　　　　　D．android:layout_ toLeftOf
4．在表格布局中，android:layout_column 属性指定(　　)。
　A．行数　　　B．列数　　　C．总行数　　　D．总列数

三、简答题

1．为什么需要布局嵌套？
2．线性布局能否实现网络功能？
3．表格布局和网格布局的相同点与不同点有哪些？

4. 绝对布局的优点和缺点有哪些？

四、实验题

1. 线性布局和相对布局（参考教材 4.1 节）。
（1）根据教材线性布局和相对布局实例操作，模拟运行测试。
（2）用线性布局和相对布局设计图书查询页面。
2. 表格布局（参考教材 4.1 节）。
（1）根据教材表格布局实例操作，模拟运行测试。
（2）用表格布局设计输入学生的基本信息。
3. 网格布局（参考教材 4.1 节）。
（1）根据教材网格布局实例操作，模拟运行测试。
（2）用网格布局显示下列界面。

4. 绝对布局和版块布局（参考教材 4.1 节）。
（1）根据教材绝对布局实例操作，模拟运行测试。
（2）根据教材版块布局实例操作，模拟运行测试。
5. 参考教材 4.2 节内容，完成【例一】：登录界面。
6. 参考教材 4.2 节内容，完成【例二】：注册界面。
7. 参考教材 4.2 节内容，完成【例三】：图书展示。

第 5 章　Android 多页面与版块

一、填空题

1. 一般采用_____和_____两种形式来使用 Intent。
2. Intent 的构成包括 Action、_____、_____、_____、_____和_____。
3. Activity 控制的页面从产生到结束，会经历_____个阶段。
4. Fragment 具有与 Activity 类似的生命周期，在_____不同。
5. Toast 的作用是_____。
6. LogCat 区域中有 V、d、i、w、e 等 5 个字母，其中 w 代表_____。

二、选择题

1. 一个 Android 应用程序，系统默认包含（　　）Activity。
A. 0 个　　　　　B. 1 个　　　　　C. 2 个　　　　　D. 多个
2. （　　）方法，Activity 从启动到关闭不会执行。

A．onCreate() B．onStart() C．onResume() D．onRestart()
3．不能使用 Intent 启动的是（ ）。
A．Activity B．服务 C．广播 D．内容提供者
4．startActivityForResult 方法接收两个参数，第一个是 Intent，第二个是（ ）。
A．resultCode B．requestCode C．请求码 D．data
5．在 Activity 的生命周期中，当 Activity 处于栈顶时，处于（ ）状态。
A．活动 B．暂停 C．停止 D．销毁
6．在 Activity 的生命周期中，当 Activity 被某个对话框覆盖掉一部分之后，处于（ ）状态。
A．活动 B．暂停 C．停止 D．销毁

三、简答题

1．Android 的三个基本组件（Activity、Service 和 Broadcast Receiver）传递 Intent 方式是什么？
2．通过 Bundle 在 Activity 之间传递数据的过程是什么？
3．基本数据类型传递方式和对象数据类型传递方式的特点有哪些？
4．Activity 的生命周期的时间点是什么？Activity 的生命周期在编程中的意义有哪些？

四、实验题

1．参考教材 5.1 节内容，完成【例一】：登录响应。
2．参考教材 5.1 节内容，【例二】：注册成功直接登录。
3．参考教材 5.2 节内容，【例三】：登录响应-生命周期。
4．参考教材 5.3 节内容，【例四】：分类预览图书。
5．参考教材 5.3 节内容，【例五】：分类预览图书-生命周期。
6．设计一个计算 $ax^2+bx+c=0$，一个页面输入系数，另一个页面分别通过基本数据类型传递方式和对象数据类型传递方式获得输入系数，计算方程的根并且显示计算结果。
7．在计算 $ax^2+bx+c=0$ 应用中加入显示生命周期提示信息。

第 6 章 Android 用户界面进阶

一、填空题

1．选项菜单可以用 XML 文件描述菜单，还可以_____描述菜单。
2．java 类文件复制到新功能，需要修改_____。
3．XML 文件描述菜单包括_____。
4．能够适应屏幕大小包含的是_____。

二、简答题

1．三种菜单的特点、使用场合有哪些？
2．代码描述菜单的优点有哪些？
3．操作栏与菜单相比有什么方便之处？
4．什么情况适合 Fragment 页面？
5．什么情况适合导航栏？

三、实验题

1. 参考教材 6.1 节内容，完成【例一】：调用第 4 章例二、例三和第 5 章例一。
2. 参考教材 6.1 节内容，【例二】：第 4 章例二、例三组和第 5 章例一分类组。
3. 参考教材 6.1 节内容，【例三】：根据第 4 章例三选择图书显示详细信息。
4. 参考教材 6.1 节内容，【例四】：实现例二分组菜单。
5. 参考教材 6.2 节内容，【例五】：图书列表和详细信息不同页和同页显示。
6. 参考教材 6.3 节内容，【例六】：实现例二分组菜单。
7. 用各种菜单方式实现酒店订餐系统。

第 7 章　Android 服务与广播程序设计

一、选择题

1. 每一次启动服务都会调用（　　）方法。
   A. onCreate()　　B. onStart()　　C. onResume()　　D. onStartCommand()
2. （　　）方法不属于 Service 生命周期。
   A. onResume()　　B. onStart()　　C. onStop()　　D. onDestroy()
3. 继承 BroadcastReceiver 会重写（　　）方法。
   A. onReceiver()　　B. onUpdate()　　C. onCreate()　　D. onStart()
4. 下列方法中，用于发送一条有序广播的方法是（　　）。
   A. startBroadcastReceiver()　　B. sendOrderedBroadcast()
   C. sendBroadcast()　　D. sendReceiver()
5. 在清单文件中，注册广播时使用的节点是（　　）。
   A. <activity>　　B. <broadcast>　　C. <receiver>　　D. <broadcastreceiver>
6. 属于绑定服务特点的是（　　）。
   A. 以 bindService()方法开启　　B. 调用者关闭后服务关闭
   C. 必须实现 ServiceConnection()　　D. 使用 stopService()方法关闭服务
7. Service 与 Activity 的共同点是（　　）。
   A. 都是四大组件之一　　B. 都有 onResume()方法
   C. 都可以被远程调用　　D. 都可以定义界面
8. 关于 Service 生命周期的 onCreate()和 onStart()方法，正确的是（　　）。
   A. 如果 Service 已经启动，将先后调用 onCreate()和 onStart()方法
   B. 当第一次启动的时候先后调用 onCreate()和 onStart()方法
   C. 当第一次启动的时候只会调用 onCreate()方法
   D. 如果 Service 已经启动，只会执行 onStart()方法，不再执行 onCreate()方法

二、简答题

1. 什么情况需要使用线程？
2. Service 的几种启动方式及其特点有哪些？
3. 多个 Service 交互的特点有哪些？
4. 什么情况需要使用广播？普通广播和有序广播应用场合有哪些？

三、实验题

1. 参考教材 7.2 节的内容，完成启动方式使用 Service 的实例。
2. 参考教材 7.2 节的内容，完成绑定方式使用 Service 的实例。
3. 参考教材 7.2 节的内容，完成多个 Service 交互及生命周期实例。
4. 参考教材 7.3 节的内容，完成普通广播应用实例。
5. 参考教材 7.3 节的内容，完成有序广播应用实例。

第 8 章 Android 数据存储与共享

一、填空题

1. SharedPreferences 本质上是一个 XML 文件，存储的数据是以_____的格式保存的。
2. SharedPreferences 用于存储应用程序的_____。
3. 指定文件内容可以追加的文件操作权限是_____。
4. 创建数据库可以通过使用_____方法创建数据库。另外还可以通过写一个继承_____类的方式创建数据库。
5. 查询 SQLite 数据库中的信息使用_____接口，使用完毕后调用_____关闭。
6. ContentProvider 对数据进行操作的方法包括_____。
7. ContentResolver 可以通过 ContentProvider 提供的_____进行数据操作。
8. ContentProvider 创建时首先会调用_____方法。

二、简答题

1. 采用什么说明方法可以减少不同的存储数据的方法对其他程序的影响？
2. Android 系统文件访问为什么需要采用两个函数打开文件？
3. 如何查看采用 SQLite 数据库方式存储的数据文件？
4. 说明 ContentProvider 的调用关系。

三、实验题

1. 参考教材 8.1 节的内容，完成【例一】：采用 SharedPreferences 方式，存取注册信息实例。
2. 参考教材 8.2 节的内容，完成【例二】：采用文件存储方式，存取注册信息实例。
3. 参考教材 8.3 节的内容，完成【例三】：采用 SQLite，存取注册信息实例。
4. 参考教材 8.4 节的内容，完成【例四】：采用 ContentProvider，获取注册信息实例。

第 9 章 Android 地图应用开发

一、简答题

1. 在工程中如何引入百度地图功能？
2. 如何设置应用的开发密钥、所需权限？
3. 说明 Poi 检索功能，为什么需要添加类库？

二、实验题

1. 参考教材 9.1 节的内容，下载百度地图包，安装百度地图应用环境。

2. 参考教材 9.2 节的内容，完成根据选择的城市和区检索，按照地图类型显示实例。
3. 参考教材 9.3 节的内容，完成根据经纬度查找地址，或者按照地址查找经纬度实例。
4. 参考教材 9.4 节的内容，完成运用 Poi 来检索南京市内的书店实例。
5. 参考教材 9.5 节的内容，完成根据输入的起始地址和到达地址规划驾驶路径实例。
6. 参考教材 9.6 节的内容，完成根据输入的城市和公交车线路编号，查询公交车沿途站点及其连线图实例。